Customer Experience 3.0

High-Profit Strategies in the Age of Techno Service

JOHN A. GOODMAN

HARPERCOLLINS
LEADERSHIP

AN IMPRINT OF HARPERCOLLINS

For bulk discounts, please visit:
https://www.harpercollinsleadership.com/bulk-sales/
Or contact us at
Email: hcleadership@harpercollins.com
Phone: 1-800-250-5308

Published by HarperCollins Leadership, an imprint of HarperCollins Focus LLC.

Any internet addresses, phone numbers, or company or product information printed
in this book are offered as a resource and are not intended in any way to be or to imply an
endorsement by HarperCollins Leadership, nor does HarperCollins Leadership vouch for the
existence, content, or services of these sites, phone numbers, companies, or products beyond the
life of this book.

ISBN 978-0-8144-3389-8 (eBook)
ISBN 978-0-8144-3388-1 (HC)
ISBN 978-1-4002-3107-2 (TP)

Library of Congress Cataloguing-in-Publication Data

Library of Congress Control Number: 2014004639

Contents

SECTION ONE

The Customer and the Implications of Customer Experience

SECTION **TWO**

Designing the End-to-End Customer Experience

SECTION **THREE**

Key Issues of Implementation

Foreword

When I worked as a manager at the Chevrolet Division of General Motors in the 1980s, the auto industry faced a series of new challenges.

- Products, which had been relatively simple, were becoming more complex.

- Owner's manuals were getting thicker, and customers were reading less of them.

- Consumers were becoming more demanding.

- Foreign competitors were improving quality, and a new process called TQM (total quality management) was challenging American manufacturers to improve product design and physical quality.

- Following the latest study commissioned by the White House Office of Consumer Affairs in the mid-1980s, both the media and the government were taking an interest in consumer protection and complaint handling.

John Goodman and his business partner, Marc Grainer, led the White House–sponsored studies of customer service and complaint handling. What they found changed the way corporate America viewed consumer affairs and customer service. Goodman and Grainer showed that when customers complained and were satisfied, they became more loyal. Strong customer service became profitable because it kept customers who would otherwise have switched brands.

The Cola-Cola Company sponsored two additional studies led by Goodman. The first found that unhappy customers told twice as many acquaintances about a bad experience as they did about a good one. The second showed that when customers were educated by the company on the content of their contract and the customer's responsibility, they encountered fewer problems and found the company easier to deal with. The result was customers who were more loyal and more willing to recommend the company or product.

These studies were reported in *BusinessWeek* in 1984 as one of the reasons that the GE Answer Center was established and American Express and General Motors initiated the use of 800 numbers to centralize their complaint-handling processes. With Goodman and Grainer's assistance, General Motors created the Buick and Chevrolet Customer Assistance Centers; I was privileged to be included on its launch team.

In the fall of 1989, I joined Toyota Motor Sales, U.S.A. Inc. and assisted the company (with Goodman's support) in enhancing both the HQ customer assistance center and the field customer service strategy. Toyota faced challenges similar to those that GM had faced in the previous decade—with some differences. Toyota's design and manufacturing quality were better than most of their competitors', and customer satisfaction was higher, but these issues brought their own challenges.

- Customer expectations were rising rapidly just as cars were evolving from a straight mechanical device to a myriad of electronic devices on wheels.

- Product functions and repairs were becoming more complex.

- As sales and market share continued to rise, the organization and the dealers were challenged to continue delivering a quality customer experience.

- Government and media scrutiny became more intense, focusing on less prevalent issues as well as those associated with customer maintenance and driving behavior.

The impact of these trends was exacerbated by several other market trends. First, customers generally failed to read owner manuals or maintenance guides. Second, computers had become a mainstay of business operations, including the customer service function. Although the computer systems provided more information to the customer service department and allowed the tracking of product and operational performance, they required a huge investment and were often inflexible because they were driven by mainframes.

The strategy of enhancing sales and positive word of mouth by delighting customers and exceeding their expectations was popularized by companies such as Nordstrom. Luckily, at Toyota, the company understood and supported the investment in service and quality, and we had a strong dealer network with visionary dealers such as Carl Sewell. Carl wrote *Customers for Life*, which described how to create highly profitable, successful businesses based on service and ongoing customer loyalty.[1]

Fast-forward to the last three years: All things technological have changed dramatically, but other market trends have remained the same. For example:

- Products are not just electronic. Many are built with full internal computers. This is a blessing but often increases complexity for customers.

- Wireless and video technology allows companies to convey huge amounts of information to customers, such as how-to videos for mobile applications.

- Many products can now communicate diagnostic trouble codes to the manufacturer that assist technicians in troubleshooting and repairs.

- Online communities, review sites, and video sharing allow consumers to quickly receive assistance and information from the public, but they can also provide a forum to vent dissatisfaction for millions of other consumers.

The company that Goodman is associated with now, Customer Care Measurement & Consulting (CCMC), has conducted several national customer rage studies that measure customer satisfaction and customer reaction to poor service. CCMC's 2013 study showed that despite all the efforts to give good service, customers are no more satisfied than in the past.[2] Further, the study found that what customers want most are explanations and apologies, neither of which costs the company money. When it comes to investment in service, we seem to miss the logic of making small investments to get big returns. *Customer Experience 3.0* spells out strategies to strengthen the customer experience and to quantify the payoff of returns on service and a positive customer experience.

In this book, Goodman covers four strategies that a company must focus on in order to maximize its customers' experience as well as its profits.

1. Companies should design the product and the marketing strategy to set and meet reasonable customer expectations—what he calls doing it right the first time. Part of this strategy is to identify possible areas of customer disappointment or areas where customers may perceive that they are not receiving full value from the product. Companies must then proactively reach out to customers to educate or even to warn them of product limitations. The best defense is a good offense.

2. Customers must be encouraged to seek assistance when they have questions or problems—a silent, unhappy customer is a less profitable customer. Companies must provide effortless communication channels for customers seeking assistance.

3. Companies need to create an empowered service system that allows employees to fully handle a problem, educate the customers on how to receive the most value from the product, and create inexpensive emotional connections.

4. Companies must build a voice of the customer process that gathers information from across the entire customer lifecycle

from multiple data sources and that integrates the process into a single, unified picture of the customer experience. To ensure impact and the secure resources needed to deliver a strong customer experience, the process must quantify the revenue and word-of-mouth impact of problems and opportunities.

For over 30 years, I have known and worked with John on numerous projects—and he rarely gives a simple answer. Instead, he challenges me by suggesting possible strategies and approaches, often outlining current best practices. He does the same in *Customer Experience 3.0*. He weaves in suggestions on how to use technology to understand, serve, and listen to the customer. His goal is to move your organization from firefighting to prevention through proactive consumer education and service—again often using technology. I have used his first book, *Strategic Customer Service*, as a guide to enhance service in the areas that I manage. I am sure you will find *Customer Experience 3.0* an excellent guide to managing both the basic customer experience and the myriad of new technologies that are now available for delivering a great customer experience.

Richard DuFresne
National Manager Customer Care, Toyota Motor Sales Corporation

Notes

1. Carl Sewell and Paul Brown, *Customers for Life* (New York: Doubleday, 1990).
2. Scott Broetzmann, Marc Grainer, and John Goodman, *2013 National Rage Study* (Alexandria, VA: Customer Care Measurement & Consulting, 2013).

Why Customer Experience 3.0?

My career began 40 years ago when I examined government and private sector complaint handling for the White House Office of Consumer Affairs. It quickly became clear that many consumer problems could be prevented if the marketing department set more realistic expectations and customers were educated on how to use products and encouraged more to read and follow directions. Many of my articles in the 1980s for quality, marketing, and customer service publications advocated for these actions, but only customer service and consumer affairs professionals listened.

Soon thereafter, authors, such as Joe Pine, Shaun Smith, and Jeanne Bliss, advanced the concept of customer experience (CE), which is dramatically broader than customer service. Customer service basically involves complaint handling, whereas CE covers everything from initial consumer awareness of the product to final use—and it requires support from the entire company. Pine, Smith, and Bliss urged companies to establish an executive position to manage CE. In the last decade, the literature on CE has grown and fragmented.

Now there are books on social media, delivering WOW service, customer satisfaction management and measurement, how to be a chief customer officer, and even how to use technology to completely eliminate the need for service. However, I have found no book on delivering *an end-to-end CE with measurable financial payoff using aggressive service, technology, and emotional connection*. The key words to leading a successful CE effort are:

- *End-to-end*—The product design and marketing teams must set proper expectations and the *whole* company must work together to meet customers' reasonable expectations and to manage their unreasonable expectations.

- *Measurable financial payoff*—If you don't have one, the chief financial officer will not accept your case for investment in an improved CE, and you will fail.

- *Aggressive service*—Barriers exist in almost every organization to customers easily seeking assistance. You must break them down, including actually encouraging complaints and questions.

- *Using technology*—Technology can be applied successfully only by following a process map of the ideal CE to transparently and proactively educate in order to prevent problems and to provide service via multiple channels.

Customer Experience 3.0 also addresses several service, marketing, and technology myths and misunderstandings. For instance:

- Customers complain most of the time when they have a problem. If you believe this, look at your own behavior.

- It costs more to give great service than to give good service. Actually, the enhanced revenue and margin payoff of good service is usually ten times the incremental cost of providing great service.

- Social media are the key to market success. In reality, the vast majority of word of mouth (WOM) is still offline and driven by basic service interactions.

- Smartphones are the new key revolutionary tool for CE. Smartphones simply speed up current interactions; what is even more revolutionary is so-called Big Data, speech analytics and wireless communication of products to companies (and vice versa).

- Websites are passé. Almost all customers go to the website before calling or tweeting; most websites are terrible from the perspective of customer service and proactive education—but companies still let the IT department run the website and the marketing department monopolize its content.

- The lack of good staff is the cause of most customer dissatisfaction. Studies show that staff who are truly empowered by supervisors and who have access to information are happy employees and effective at providing positive CE.

Speaking of positive CE, let me step you through the book, which is divided into three sections.

Section One: The Customer and the Implications of Customer Experience

This first section of the book describes how to set customer expectations and the implications for managing the customer's experience, future loyalty, and the word-of-mouth impact of a great CE. My research on customer behavior includes customer reaction to a problem, word of mouth, and word of mouse (the Web). I present an experience framework for setting and fulfilling those expectations. Finally, this section explains how your efforts to understand, set, manage, and fulfill customer expectations can be translated into a measurable financial impact that the chief financial officer and the chief marketing officer will embrace.

Section Two: Designing the End-to-End Customer Experience

The second section of the book describes how to create the four parts of the end-to-end CE framework: (1) doing it right the first time, (2) providing access to service, (3) providing caring service, and (4) listening and learning. If you cannot measure it, you cannot manage it. Therefore, process and outcome metrics are suggested in each of the four parts of the framework.

The first step is mapping the existing CE and the ideal CE to ensure you are not building unpleasant surprises into the CE. Along with mapping CE, you need to create a customer-focused company culture in which the customer and the employee become partners in the experience. In addition, I discuss psychic pizza—anticipatory service and communication. Using Big Data, a company can determine what customers' next needs will be and use technology to deliver the service or information before customers know they want or need your help.

The second step to having an end-to-end CE framework is providing access to service. This step is highly counterintuitive; most executives believe companies should do everything possible to reduce complaints and contacts. I advocate begging the customer to contact the company with the message, "We can solve only the problems we know about!" Many company executives frown on this concept, until you show how much money is left on the table when you do not hear from unhappy customers. Creative approaches to pushing customers to communicate are presented. I also address the challenges of using multiple communication channels.

The section addressing the third step—providing caring service—also advocates for a dramatically expanded role for the service function and frontline service staff. In the past, frontline staff simply had to handle the customer's problems and hoped they achieved resolution on first contact. I recommend going further by training the frontline staff to also educate the customer on how to receive more value from the product. In addition, the staff should be explicitly responsible for creating an emotional connection with the customer, when appropriate, and gather information for use in the Voice of the Customer (VOC) process.

The final step, listening and learning, builds an end-to-end VOC process and ensures that the output has a constructive impact across the company and a positive impact on the customer. The three challenges addressed in this final step are: (1) creating an end-to-end VOC process that includes multiple data sources; (2) integrating the multiple data sources into a unified picture of the CE; and (3) packaging the VOC data with a financial analysis that creates the economic imperative for company action. If you quantify the cost of inaction, you precipitate action.

Section Three: Key Issues of Implementation

The final section of the book focuses on the vital issues that have, in my opinion, led to the failure or dilution of most CE efforts: the improper use of technology, not empowering employees to be flexible in handling a customer's problem, and the lack of savvy CE leadership.

Technology has revolutionized CE for both the customer and the company. Everything, including faucets, now contains microchips, and cars are functioning as computers on wheels. Complexity has increased, but as Rick DuFresne noted in his foreword to this book, nobody reads directions anymore. Further, most CE executives delegate too much process decision making to the technology department. At the same time, technology executives are often the first to admit that they are not experts in customer service or in setting proper customer expectations. This section shows how a partnership between CE and the proper management of technology can have huge positive impacts.

The section on empowerment addresses how a company can foster flexibility and risk taking and, most importantly, train supervisors to embrace empowering employees. Empowerment is important because it is the key to emotional connection with the customer and allows resolution of the problem and the extra time and effort for connection. Connection seldom happens if an employee cannot resolve a customer's basic request. Therefore, empowerment is a prerequisite to connection. Because events and customers are unpredictable, true empowerment must allow the employee to be flexible and to take risks. I also discuss inexpensive methods of creating emotional connection; it need not be time-consuming and costly. Finally, several counterintuitive metrics are suggested for measuring whether true empowerment and connection are taking place.

In my experience, CE executives who are not successful are those who cannot justify their bottom-line contribution to corporate management. The secret to success is to act as an internal consultant who makes the internal client look good and demonstrate a positive bottom line impact as well. This section of the book suggests the appropriate roles for both the CE executive and for the non–executive director of CE. I then share best practices for getting started and sustaining positive impact. Finally, I suggest metrics for evaluating the CE function.

In many areas of the book, I raise a range of issues and scratch the surface of broad topics. I then focus on the best practices, key pitfalls to avoid, and the best metrics for measuring the impact of the different parts of the CE process. My intention for this book, like my previous book *Strategic Customer Service*, is to provide a practical guide for getting things done. I also demonstrate the impact of CE on the bottom line, based on simple, easily measured metrics. I have used the methods described in every type of business, nonprofit organization, and government agency. I hope the book's logic and simplicity will help you reduce the frustration of both customers and employees, while enhancing your bottom line.

This book is a culmination of four decades of research, and there were many important contributors. The principals of CCMC, Scott Broetzmann and Marc Grainer, whom I have worked with for 40 years, contributed much to the substance of this book. Brad Cleveland, founding partner of the International Customer Management Institute, and Peter W. North, CEO of True North, contributed significantly in the service access and technology/social media areas, respectively.

David Beinhacker, CCMC Director of Research, contributed data to bolster many of my assertions. Data and advice on international customer experience was provided by Nobu Hatanaka of Learningit in Tokyo, Japan, and Eduardo Laveglia of Proaxion in Buenos Aires, Argentina.

Richard DuFresne of Toyota provided many real-world observations as well as the foreword to this book. Lynn Holmgrem, who has led customer experience at Whirlpool, Frontier Communications, and Millicom Cellular, implemented many of my concepts and allowed me to report their impact. Many of my former associates at TARP also contributed insights, especially Cindy Grimm and Crystal Collier.

My wife, Alice Goodman, applied her decades of experience writing congressional testimony to do a masterful job of making this book clear and concise. Libby Koponen was an excellent editor in polishing the manuscript, as well as questioning issues that would not be clear to non–customer service veterans. Michael Snell is the most perfect, savvy agent an author could want and also was a cheerleader when things got difficult. Bob Nirkind of AMACOM was even more patient with me while I was writing this book than with the last one. His guidance kept me on track and focused.

If at any point you find that something is not working or you believe I have given bad advice, I practice what I preach. Please let me know—I can solve only the problems I know about—and I will do what I can to make you successful. Feel free to contact me at jgoodman@customercaremc.com.

1

The Customer and the Implications of Customer Experience

CHAPTER 1

Why Good Service Might Not Result in a Great Experience

My wife bought me a high-end brand-name surround sound system for my birthday. Recently, the four external speakers stopped working. After much fiddling, using the poorly written manual as a guide, I called the manufacturer's 800 number, navigated the phone menus as best I could, and waited 20 minutes before giving up. Two days later, I tried again, and again gave up on a long queue. Finally, I arranged for the local retailer to repair the system onsite. The technician was equally stumped and called the manufacturer, only to learn that the unit needed to be sent in for repair. Many weeks later, the company returned my system. The manufacturer's support center most likely recorded a successful once-and-done call. However, every time that brand is mentioned, I wince and want to tell my story to save someone else from making the same mistake.

Chances are high that your organization's overall customer experience (CE) is worse than you think. I refer to customer experience rather than customer service because service is only one piece of the experience. Most executives think that if their staff is courteous, responsive, and effective, their company is delivering a great CE. Not true—because CE is end to end, ranging from honesty in marketing through the product lasting beyond the warranty period. This lack of differentiation between CE and service is the first of a myriad of mis-perceptions that prevent executives from getting an accurate under-standing of their current level of CE. This also precludes executives from understanding the huge potential payoff of enhancing the CE.

The biggest challenge for providing a great CE is to pull yourself and your business out of complacency. Most executives are compla-cent about the CE they are providing, unaware of the serious damage to their operations and their bottom lines.

This chapter is intended to jolt you out of that complacency. The five most important realities of CE, many of which will surprise you, are identified and discussed:

1. Customer expectations are simpler to understand and harder to fulfill than most people think they are.

2. Employees do not cause most customer dissatisfaction.

3. No news is not necessarily good news; a company receiving few complaints is not always delivering a great CE.

4. Your company's current CE is needlessly leaving huge amounts of money on the table that could easily be added to the top and bottom lines.

5. Technology has changed what customers expect from a com-pany and how the company delivers the CE; most executives do not realize how inexpensive technology is when used effectively.

In addition, I explain how to use each of the five realities to your competitive advantage.

Understanding Customer Expectations

Customer expectations for any experience are seemingly very simple: Customers do not want unpleasant surprises; they expect easy access to service on their terms; they want first-time resolution to requests for assistance; and they require evidence that the company cares. However, meeting these expectations is more complex than it may seem.

No Unpleasant Surprises

For most products and services, customers do not expect a WOW experience. But they do expect what is promised and what they ordered without any hassles or unpleasant surprises.

The first part of delivering what you promise is making sure the customer understands what was pledged. This requires a company's marketing to be honest and customers to read and understand the contract between them and the company on what is to be delivered. The challenge here is that not many people read descriptions, directions, or contracts, especially anything with fine print or footnotes. Executives must understand that the inclusion of any fine print is building unpleasant surprises into the product or service.

The second part of delivering what you promise is doing it flawlessly and memorably. Think about your last airline flight. If the flight took off and landed on time along with your baggage, you probably barely remember the details unless you had personal interaction with an employee. Airlines usually differentiate themselves only when problems occur: How well did they handle the situation? Much of Chapter 6 will focus on how to make problem handling memorable in a positive way.

Part of delivering a great CE is anticipating possible sources of dissatisfaction or uncertainty and addressing the sources before they occur. This can be as simple as the cable company confirming that the technician is actually coming tomorrow between 8 A.M. and 9 A.M., or the pilot announcing a 30-minute air traffic delay. Ideally, you deliver the answer to the question five minutes *before* the customer asks it. I call this psychic pizza—as though the deliveryperson rings your doorbell and says, "Here's the pizza you were about to order!" There will be much more on anticipatory service later in Chapter 4.

Easy Access to the Service System

Unless companies make it easy for customers to complain, most will not bother; they will simply take their business elsewhere. Thus, companies need to both encourage customers to contact them and make it easy to do so.

The first step is counteracting the common belief that companies do not want to hear about problems. Make it clear that you genuinely want to help by sending the strong message that the company can solve only the problems that it hears about.

Further, the actual act of complaining must be effortless, or customers will not bother to provide feedback. This is especially true of twenty-somethings, who complain less than older people. In most cases, twenty-somethings will not even call an 800 number; they would rather go to the website or Google the solution. Information should be easy to find without calling the company. Ideally, this information should be provided where and when the customer needs it, through whatever channel (phone, Web, text) is most appealing to the customer.

The importance of this effortlessness can be seen in the following story. A soft drink company in Europe shipped drinks that tasted a bit off. Unfortunately, their hotline did not accept texts, even though young people were their biggest customers, and it was open only from 9 A.M. to 5 P.M., Monday through Friday. Their customers drank soft drinks most in the evening and on weekends—but who is going to save an empty can on Saturday and then call on Monday? Because the company accepted phone calls only during business hours, it did not realize the magnitude of what was happening until the problem reached disastrous proportions. They ultimately removed the product from shelves across much of Europe at a cost of hundreds of millions of Euros. So make sure help is available to the customer when the product or service is most likely to be used or when they are thinking about your product or service (e.g., reading a monthly billing statement).

Also, effortlessness involves setting up a communication channel that is easy to use. A complicated, multilayered phone tree (technically called an interactive voice response, or IVR, system) is a huge barrier. On the other hand, when the 800 number listing is displayed along with the three or four simply worded options included in the first level of the menu (for example, "Speak to a live rep"), the success rate will rise significantly. Twenty percent more callers will complete their calls, and satisfaction will be much higher because customers do not need to listen for the right option; they already know which option to push before they dial the number.

A website with a mass of information or hundreds of frequently asked questions (FAQs) also makes it difficult and time-consuming to find the right answer. A home page with the five most common questions will probably answer a majority of the major concerns with one click.

The best communication channels facilitate response to the customer's request immediately—such as the home page with the list of the five most common questions—or with only a minor wait in a queue, usually less than one minute on the phone. This standard requires coordination between the access channels such as the webpage, phone lines, and the service response process to guarantee that the service system has the capacity to handle the workload. For example, an easily accessed 800 number is useless if no one answers or has the information customers need.

Chapter 5 will address the details of developing a successful strategy for ensuring that all the different types of customers you serve will gain easy access to service via their preferred channels.

First-Time Resolution to Requests for Assistance

Customers want to feel they are being treated fairly and effectively, and part of achieving this is resolving their requests for assistance. You can make customers feel fairly and well treated by:

- Addressing all major aspects of the request with your first response or in a timely manner.

- Fully meeting the customer's expectation. The prevalent, incorrect myth that companies should not just meet but continuously exceed expectations—often a wasteful, misguided premise—will be discussed in Chapter 4. Additionally, you must recognize that not all customers can have all of their expectations met. Often the answer "no" is appropriate and acceptable if accompanied by a clear explanation. This will be addressed shortly.

- Explaining the reason for the remedy or action clearly and without jargon.

These actions apply to all responses, whether they are face to face, by telephone, or through a technological self-service system. Keep in mind that the very fact that a problem has occurred often causes damage to the relationship—unless service and recovery are stellar. If incorrect expectations are set up front, trust and loyalty may be lost for good.

It is a fact that good service does not necessarily equate to a great overall CE. Often the customer correctly feels that the problem should never have occurred in the first place.

Evidence That the Company Cares

When unmet expectations cannot be rectified, the only action company employees can take is to empathize and show that they care. Never underestimate the importance of empathy; most customers accept that "stuff happens." When it does, they want sympathy and acknowledgment that the company feels their pain. A heartfelt apology can do wonders. It is even better if the company goes beyond that and also provides some tangible token compensation.

An example is when a customer misses a flight connection or when mechanical problems with the airplane result in a flight cancellation and a long night in an airport. The worst approach is to legalistically deny all responsibility. In these situations, the most an airline can do is quickly provide some physical token comforts (e.g., food and a beverage) and show that it understands and cares. A company can effectively show that it cares by:

- Establishing rapport by paraphrasing what the customer has said, also known as reflective listening, and using a sympathetic, understanding tone of voice. The right tone (even in a chat or email) can convey concern; the repeated words show the customer he or she has been heard.

- Exhibiting empathy, including sometimes offering an apology even if the company does not accept blame for causing the problem.

- Creating an emotional connection, which involves acknowledging a common experience that conveys, "I understand and sympathize with your pain."

- Listening lets the customer know he or she has been heard. People feel heard when the company representative takes reasonable action and assures customers that their issues will be used to improve the company's offerings.

Identifying Sources of Customer Dissatisfaction and Uncertainty

Executives assume that dissatisfaction is almost all due to individual employee behavior, either their actions or attitude. However, ask yourself how many of your employees actually come to work thinking, "I'm going to intentionally provide bad service to most of the customers I encounter today." Almost everyone wants to be successful.

Although some dissatisfaction may emanate from employees' actions, the majority of dissatisfaction occurs when the product is delivered exactly according to spec! These causes are rooted in both the customer and the company. Customers misuse products, and companies poorly design, market, and produce products and services. In other words, the product can be delivered exactly as intended and still result in dissatisfied customers because they feel misled by the product description or marketing, or simply do not like the product. No one in the customer service area is responsible for customer expectations or the actual product, but often customer service takes the brunt of customers' anger.

Ironically, issues that leave the customer feeling misled—most often, marketing and sales communications—cause more dissatisfaction and damage to loyalty on a per-problem basis than when a product or service is actually defective. This is because the customer believes he or she has been intentionally misled. Unfulfilled expectations due to what the customer feels is deceit often create two to four times more damage to loyalty than a simple frontline staff mistake or broken product.

Across all industries, the company itself is the primary source of dissatisfaction and unpleasant surprises. Based on reviews across hundreds of companies, I estimate that the distribution of causes across the three broad categories are:

- Customer caused: 20–30 percent.

- Company caused: 40–60 percent.

- Employee attitude, error, or failure to follow policy: 20–30 percent.

Customer-Caused Dissatisfaction

Customers seldom read directions and often make assumptions (some reasonable and others unreasonable) about how a product should operate. As a result, customers often misuse the product. For example, a liquid bleach company routinely receives suggestions to improve the bleach's taste from consumers who use it to whiten their teeth (not a recommended use!). In an attempt to use humor to get customers to help themselves, one consumer electronics manufacturer includes a card in the top of its boxes that says, "When all else fails, try reading the directions."

I have asked dozens of audiences if they read their auto owner's manuals or homeowner's insurance policies. Consistently, fewer than 2 percent said yes. Given this customer behavior, companies must simplify products so that customers do not make errors in using them.

Company-Caused Dissatisfaction

Companies create customer problems by building unpleasant surprises into products, by setting incorrect expectations among customers via misleading marketing and sales, by creating disappointment via processes that are broken or have disconnects between functional units.

Product Design

Products that are not designed intuitively lead to frustration and error. Overengineering and complexity lead to failure. For example, most consumers use only a few of the 30-plus buttons on their TV remotes and become frustrated if they accidently push the wrong button and do not know how to recover. Two solutions to this problem are product simplification (e.g., putting fewer buttons on the remote) or highlighting the most frequently used buttons (e.g., making the on/off, channel, mute, and volume buttons larger and a different color).

Misleading Sales and Marketing

Misleading marketing and sales materials do the most damage on a per-problem basis because customers feel they have been intentionally misled. Customers are willing to forgive manufacturing or operational mistakes, but they will not forgive being intentionally misled. When the sales or marketing departments fail to clearly communicate product limitations, they are creating dissatisfaction of the worst kind.

The best practice to avoid misleading customers is to create a simple product or to warn the customer about potential misunderstanding. For example, an insurance company overtly warns the customer about exclusions by highlighting the maximum amount of jewelry and firearms covered in its homeowner's policy. Although its sales department was nervous about highlighting limitations, customers were pleased to be informed and almost always bought a rider for insurance at a higher premium.

Any contract, warranty, or financial service account with fine print or nonintuitive definitions and footnotes is open to creating dissatisfaction. For instance, most automobile tire warranties cover road hazards but not nails or cuts to the side of the tire that result from a rock or pothole. Consumers expect that these damages are covered due to their understanding of the term road hazard and are left unsatisfied upon learning that they must pay for the damage. A warranty that says "bumper to bumper or lifetime" and then hides the exclusions or limitations in the fine print is another common example.

Broken Processes

Broken processes across company silos are another major contributor to customer dissatisfaction. For example, at an appliance company, phone reps making appointments were encouraged to rush the customer off the phone to save a minute of phone rep time (estimated as costing 50¢). The haste resulted in repair technicians not having the necessary part on the truck and incurring a second $84 technician visit to complete the repair. Later, when the phone reps were allowed two extra minutes (costing less than a dollar) to ask questions, such as, "at what point in the cycle did the dishwasher leak and from which corner," second repair visits went down 30 percent. By spending an additional dollar on phone talk time, the company saved $84 per incident and reduced customer frustration of staying home a second time waiting for a technician.

Employee Attitude or Error

Employees do make mistakes, do have bad work ethics, and on occasion do have bad attitudes. Some people are uncomfortable in customer contact positions and are probably the wrong type for the job. In these cases, it is fair to blame the employee for the poor CE. However, many other cases of alleged bad attitude, lack of knowledge, or incompetence actually stem from poor company processes, including corporate training, motivation, or systems support.

I have found it very enlightening to ask frontline staff whether they have ever been rude or unresponsive to a customer. When I preface the question with "If it happened, you probably lacked tools or support," people almost always admit to bad behavior and explain why it occurred. One airline gate agent told me, "The Operations Department would not tell me when or if the plane was going to be repaired, and when the twentieth customer asked me if the flight was still going, I forgot my training and snapped because I was embarrassed I did not know." When I quoted the agent in an article, a local businessman left the irate message, "Any rude employee should be fired on the spot!" Not necessarily. That can be a huge waste of good employees. Sometimes, a little training, or information support reinforcement is all that is needed.

No News Is Not Necessarily Good News

Companies are not aware of the number of customers who are dissatisfied because relatively few customers with problems complain. Instead of complaining, most people become less loyal and spread negative word of mouth. No news is not necessarily good news! You must understand the so-called iceberg effect, that is, the problems that companies do not hear about from customers cause at least five times as much damage as those they do hear about.

In a recent seminar exercise, the head of service for a major technology company described an occasional failure that impeded some PC operations. He was sure that almost every customer had complained because the computer glitch took place during the warranty period. He was shocked when I said my research suggested that, at most, only one-third and, more likely, one-twelfth of customers had called the company about the problem.

The vast majority of customers do not complain about problems because:

- Complaining is too much of a hassle. It is easier to live with the damaged product than work to get it fixed.

- Customers feel that complaining will do no good. They assume from previous experience in other markets that the company does not care and will not act. This is what I call trained hopelessness.

- There is a fear of retribution from the company staff. The doctor's office will not give complainers future appointments or the bank teller will do damage to their bank account.

- Customers do not know where to complain. Customers seldom take the time to research where to complain. If a customer in a retail shop cannot spot a manager or if a product does not display a phone number or email address for complaints, most people give up.

Figure 1-1 shows the iceberg effect. This finding is from the first White House–sponsored study of complaint handling.[1] In 1979 I codirected this comprehensive review of customer service in federal, state, and local government and nonprofit agencies, as well as 100 highly regarded companies. The study also included a cross-section survey of the U.S. population to identify complaint behavior when customers encountered a problem. CCMC's *2013 National Rage Study* shows that the findings are still relevant and that not much has changed.[2]

As the graphic shows, company headquarters usually hears from only 1–5 percent of customers who have encountered a problem. This small percentage means that if a company receives 10 complaints a month, 200 to 1,000 other customers may well have the same problem but are not telling you. As you will see in Chapter 3, these unarticulated complaints and unfulfilled expectations are definitely hurting your bottom line and may spell financial disaster.

Figure 1-1

Iceberg of Problems vs. Complaints Submitted to HQ

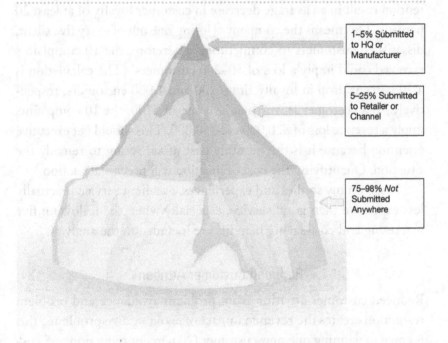

1–5% Submitted to HQ or Manufacturer

5–25% Submitted to Retailer or Channel

75–98% *Not* Submitted Anywhere

Why Your Current CE Is Leaving Huge Amounts of Money on the Table

Greed is a good way to motivate executives to improve the CE! You can use the desire to improve the bottom line to encourage chief financial officers (CFOs) and chief marketing officers (CMOs) to be CE advocates. The economic case for CE should be in clear, conservative, data-backed terms, such as: "Unarticulated, unfulfilled customer expectations are causing huge revenue losses that can be quantified in the range of $x million and recouped if we take action."

In the nearby example, the ten customers who complain directly to the company will at least provide the organization with an opportunity to resolve the problem. The majority of other 250-plus customers will not complain to anyone, and a minority will complain only to one of its distributors (also called channels) or retailers.

Based on over 1,000 separate studies I have managed, across all industries, complaints that do not come directly to the company's attention result in an average decrease in customer loyalty of at least 20 percent. This means the company is losing one out of every five silent, dissatisfied customers to competitors. Therefore, the 10 complaints received could imply a loss of 50–200 customers! (The calculation is a 20 percent drop in loyalty times 200 and 1,000 customers, respectively.) If a customer is worth an average of $100, the 10 complaints imply a revenue loss of $20,000 to $250,000. This should get executive attention because it is the monthly cost of *not* acting to remedy the situation. Quantifying the cost of inaction will precipitate action.[3]

Based on my studies and experiences, excellent service is actually less expensive than good service, especially when the following five marketing and cost-saving benefits are included in the analysis.

Reduced Customer Attrition

Reduced customer attrition from problem avoidance and problem resolution creates the revenue impact of avoiding five problems; this is equal to winning one new customer (20 percent reduction × 5 customers = one lost customer avoided). Resolving an existing problem (rather than leaving it unresolved) raises loyalty from 30 to 50 percent. The cost of winning a new customer is almost always 5 to 10 times the cost of handling a problem to a successful conclusion. In a business-to-business environment where each customer can be worth thousands or millions in revenue, it can easily cost 10–20 times as much to win a new customer as to resolve the problem.

Increased Word of Mouth (WOM)

Word of mouth (WOM) is the most cost-effective source of new customers you can have. Your customers are selling your product at no cost to you. When I ask CMOs what percentage of new customers come from referrals, the lowest estimate I hear is 20 percent; for successful companies, it can easily be 60–70 percent.

A negative experience usually causes two to four times as much negative WOM (or word of mouse[4] on the Web) as a positive experience. Great service can foster positive WOM, which reduces your marketing expenses. Once this calculation is understood, most CMOs will invest in service to foster positive WOM. Every interaction, whether sales, information seeking, or service, has the potential to produce WOM. If only 20 percent of interactions are unsatisfactory, the net stream of WOM referrals will still be negative. I have found that challenging CMOs on whether they get a higher return on investment (ROI) from advertising or from WOM based on a great CE is a very effective method of gaining CMO support for investment in the CE.[5]

Bigger Margins

Fewer problem customer experiences lead to higher margins, which, in turn, beget higher profits. Customer sensitivity to price will increase significantly when problems occur, often doubling on the initial occurrence and doubling again with multiple problem occurrences.

Lower Service Costs

Most companies could prevent 30 percent of service contacts with better communication up front. Proactive customer education and an effective Voice of the Customer (VOC) process that identifies preventable customer problems can reduce or eliminate many service contacts. The VOC process can also flag defective service processes that result in frustrated customers and employees.

Reduced Legal and Regulatory Costs

Most legal, regulatory, and public relations (PR) disasters begin as unresolved complaints. Improving CE reduces the costs from these disasters by at least 10 percent and in some cases by as much as 20 percent.

Without CFO buy-in to the economic case for a great CE, you will lose the battle. In a recent survey I conducted of 160 companies, we found that when the CFO bought into the case for a great CE, the company was five times as likely to get global customer issues fixed quickly and twice as likely to have year-over-year sustained increases in customer satisfaction.[6]

If you hope to get the resources to create a great CE, the CFO must see the opportunities to enhance the bottom line. Chapter 3 will equip you to quantify the financial opportunities of a great CE and the damage done by inaction.

Leveraging Technology for a Great Customer Experience

The right approach to CE—proactive and end-to-end management—is now highly dependent on technology. The technology available includes tools for tracking, analyzing, and acting upon customer and transaction data; managing telephone and Web interactions; and exploiting wireless communication on the business side as well as smartphones and social media to help consumers manage their whole lives. These tools, used intelligently, can easily help reduce unpleasant surprises, increase effective response, provide first-time resolution of service problems, and show that you care, all the while reducing costs.

The list that follows briefly identifies the nature, purpose, and benefit of each tool. Later chapters will address their use in more detail.

- Enterprise database technology and Customer Relationship Management (CRM), often together called Big Data, can tie operations and transaction information to customer records. This means the company knows when something has happened (customer has returned a product due to the wrong size being shipped) or is about to happen to the customer (a delivery date is going to slip or an electric bill is going to be unusually high), as well as how much the customer is worth to you. If technology is used correctly, customers will spend less time retelling their story because you already know a lot about their history.

- Telephone call management systems allow companies to manage the access (greetings, routing, and queues) to the call center and get customers to the right place faster.

- Contact management systems allow company representatives to enhance service because they know the customers' history, take appropriate action, record their actions, and often dispatch surveys after the call is handled.

- IVR systems allow customers to select from a myriad of options. These devices also collect information such as account numbers, though they are often confusing and annoying to customers as well. When properly designed, IVRs can more rapidly get customers to the right place.

- Speech and text analytics are the technologies, first used in the intelligence community, to interpret call and email content. It now can be used to evaluate, categorize, and analyze call and email records for content and emotion, thereby enhancing the impact of the VOC.

- Wireless communication now allows products to phone home, so to speak, reducing the need for physical customer intervention. For example, an office copier can automatically call the service department to request a new toner package or service visit, and a refrigerator can call a car's central command system or cell phone to remind the driver to buy milk because the carton on the shelf is almost empty. Likewise, a person can call their house security system to turn off the alarm to allow the housecleaning service to enter.

- Smartphones, with their accompanying texting, search, and email, are replacing telephone and face-to-face service to the degree that calls to 800 numbers are actually decreasing at most companies while email, text, chat, and Web service (and, to a lesser degree, social media) is increasing rapidly. Customers can get service when and how they want.

- Social media sites and the tools that monitor the sites allow both consumers and businesses to track what customers are doing and their satisfaction with each experience. Customers are better equipped with unbiased information and feel more empowered. The challenge is for businesses to act on that knowledge without looking as though they are stalking the customer.

Eliminating Unpleasant Surprises

The following case is a good example of proactively eliminating unpleasant surprises to customers. After only one month in the app store, the Mailbox app for iPhone/iPod/iPad had over 1.3 million customers who wanted to become users. Because of the nature of the app, which processes over 60 million daily emails and requires substantial computer servers to perform well, Mailbox was unable to accommodate the massive demand for its new service. Letting all users onto the service would have created a system-wide slowdown for everyone, creating an unpleasant surprise. Alternatively, flatly turning away a large portion of the would-be customers would create a more serious unpleasant surprise. The answer was to slowly add customers a few thousand at a time without overwhelming the capacity of the technology. The question then became how to explain the unpleasant surprise of the delay in receiving the service and manage the reservation queue in a way that showed empathy and caring.

To manage expectations, as well as to husband the capacity of their own computer systems, the Mailbox team allowed users to download the app and make a "reservation" to use their service. Those users who made a reservation would see a countdown that showed their place in line, as well as real-time feedback showing how many accounts Mailbox was opening per second 24 hours a day. The

ments in both technology and culture across the end-to-end CE will have dramatic payoffs. This requires three actions:

1. Can you show that the current CE is not as great as top management thinks it is? Talk with your field staff, sales reps, and retailers, and ask how many problems, grumbles, and gripes they are hearing from customers about problems that are fixable but have been dragging on for months or years. Compare the formal numbers of complaints in your systems and surveys to the numbers you hear from your conversations with the front line.

2. Is the cost of inaction greater than executives think? Identify and talk with a few lost customers about their experiences and record their case studies, ideally via video. Quantify the attrition among your customer base; normally, half of all attrition is voluntary and due to poor CE. Multiply that percentage by the size of the customer base by your best estimate of the revenue value of the average customer. Your result will show the added value of effective CE.

3. Are problems designed into the product and process? For some of the problems you identify, ask your employees about the source, arming them with the framework of sources of dissatisfaction mentioned earlier in this chapter. You will see that many wounds, if not most, are self-inflicted via process and product design and by misleading marketing and sales.

KEY TAKEAWAYS

- Customers expect to receive what is promised without unpleasant surprises, which is harder to deliver than most companies believe.

- Most customer dissatisfaction is not caused by employees but by intentional company actions and by customers who fail to read instructions, manuals, and contracts but who still blame you for the problem.

- No news is not necessarily good news; receiving few complaints does not mean you are delivering a great CE. You hear from only a small percentage of unhappy customers, which results in both complacency and the inability to recover via great service.

- Good service does not mean a great experience. The damage is often done by unpleasant surprises, and the revenue is often unrecoverable by the time service even gets involved.

- Your current CE is leaving huge amounts of money on the table, which can be quantified to motivate your CFO and CMO to invest in an improved CE.

- Technology, properly managed, can almost always make the experience effortless and memorable (via proactive service) for the customer and can be inexpensive for the company.

Notes

1. John Goodman, Marc Grainer, and Arlene Malech, *Consumer Complaint Handling in America: Summary of Findings and Recommendations* (Published under contract HEW-OS-74-292, TARP, September 1979).

2. Scott Broetzmann, Marc Grainer, and John Goodman, *2013 National Rage Study* (Arlington, VA: Customer Care Measurement & Consulting, 2013).

3. John Goodman, "The Truth According to TARP," *Competitive Advantage* (Milwaukee, WI: Service Quality Division newsletter, American Society for Quality, 1999).

4. John Goodman coined the term *word of mouse* in 1999 in the first study of online support (Cindy Grimm and John Goodman, *Industry Customer Support Benchmarking Study* [Washington, DC: Consumer Electronics Association, 1999]).

5. John Goodman, "Treat Your Customers as Your Prime Media Rep," *BrandWeek*, September 12, 2005, p. 16.

6. John Goodman and Cindy Grimm, "Factors Leading to Effective Customer Experience," *Call Center Pipeline*, January 2012.

7. Thomas Peters and Robert Waterman Jr., *In Search of Excellence* (New York: Harper & Row, 1982).

8. Tom Peters, "Push One, Push Two, Push Three—Push Your Customers Over the Edge!" *On Achieving Excellence*, 9(6), June 1994.

CHAPTER 2

More Than People

Customer Experience = People + Process + Technology

I recently wanted to upgrade several international flights with a United Airlines Star Alliance Partner. The elite service desk could not guarantee an upgrade unless I bought a very expensive nonrefundable coach ticket first; they said they needed an already purchased ticket to check availability on the other airlines and even then could not guarantee an upgrade. I asked why there wasn't a simple electronic link into the other airlines' upgrade databases and was told that United Airlines employees had been requesting that link for years. Not having the link frustrated both customers and employees.

Most executives define customer experience as an intelligent, courteous person providing personalized service. However, there is a lot more to both CE and what is needed to deliver it right. The first challenge is that CE is much more than providing service. The second challenge is that people cannot deliver a great end-to-end experience unless they are supported by well-designed processes and technology.

In this chapter, you will learn how to:

- Consistently deliver a great CE within a four-part framework
- Apply technology during each phase of the framework

- Determine the metrics needed to measure and manage CE effectiveness.

Delivering a Great Customer Experience: A Four-Part Framework

The CE covers every activity that the organization performs, including activities that do not directly touch the customer but that affect the customer's overall experience. For instance, the human resources (HR) department does not directly interact with customers, but its approach to selecting, training, compensating, and motivating employees has a huge impact on the CE. If the HR department chooses to skimp on its investment in training or on paying an above-average wage that attracts superior employees, the result can be a systematically poor experience from those employees who directly deliver service to customers.

The following four-part framework shows how all organizational activities impact the CE and help ensure that CE is designed into all organizational activities, resulting in a complete, consistent, continuously improving experience—with less need for expensive, unnecessary service. The four parts are:

- *Doing It Right the First Time (DIRFT)*—Many CE problems stem from unmet expectations. To deliver great CE, companies need to set expectations about the product accurately and give employees the tools, processes, and partners to deliver what was promised. This includes planning for CE when designing, selling, delivering, and billing for the product.

- *Encouraging easy access to service via effortless channels*—Customers with questions or problems need easy access to service. As noted in Chapter 1, many customers do not believe the company will actually help them and do not request assistance. In addition, before they can get help, customers are often forced to run a technological gauntlet of passwords, phone trees, and queues.

- *Creating a complete customer service experience via every communication channel*—A good service system is not just satisfying; it efficiently creates emotional connection, prevents future problems, and fosters positive WOM.

- *Listening and learning by creating an effective VOC process*—The final phase of delivering great CE comes from collecting feedback during DIRFT, access, and service and from using that feedback to improve the entire process. This continuous improvement creates added value and ensures that you are doing the *right job* right the first time.

Figure 2-1 summarizes the four-part framework.

This chapter describes each part of the framework, along with the roles people, processes, and technology can play in each. Later chapters explain the detailed implementation of each part.

Figure 2-1

Four Components of the Customer Experience Strategy

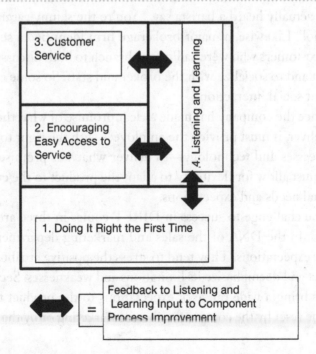

Doing It Right the First Time

When customers purchase a product or service, they expect whatever was promised by the company to be delivered with no unpleasant surprises. Most customers are not looking for their expectations to be exceeded, what everyone now calls WOWed. When you travel on an airline, you want to get from point A to B on time with your luggage. The transaction can be differentiated from the competition via value-adds, such as free drinks in first class or an enhanced social experience, but the most important thing is to get passengers to their destinations on time with their luggage.

Customers start with a set of basic expectations. At a coffee shop, customers usually want the basic transaction—a good cup of coffee and a comfortable place to drink it, often while talking with friends. Although the coffee shop is not the social experience of the day, social experience can be a designed-in component of the brand promise, even for a cup of coffee. Starbucks' customers pay a significant extra margin for a great CE, including wifi and perhaps even a barista who knows them by name or at least remembers what they usually order. I have actually heard a barista say, "You're the skinny caramel Frappuccino!" Likewise, a major brokerage firm identified a segment of older customers who were calling the branch to both discuss the stock market and to socialize with the broker and staff. So some customers do want social interactions.

Once the company has made a clear promise of what the product will deliver, it must provide the employees with the right tools—that is, processes and technology—to deliver what was promised. These tools must allow for flexibility to adapt the product to the customer's personal needs and expectations.

The challenge to success in DIRFT comes in three areas. First, it is not in the DNA of the sales and marketing departments to set proper expectations. They tend to stress the positive attributes of the product and deemphasize its limitations and weaknesses. Second, customers bring a wide range of expectations to the product that must often be reset by the company, either in marketing or by the frontline employee at the time of delivery. Third, outside factors (think weather or mechanical breakdown) and the customer's status or personal situation can all vary significantly, even for the same individual customer. A flight cancellation on a vacation trip will not tend to upset me as much as a cancellation on a flight to a business meeting when I was cutting it close to begin with. When I am on vacation, getting rebooked on the next flight would probably be satisfactory. When I am going to a business meeting, an airline wins points by proactively rerouting me through another city so that I can arrive more or less on time. Employees and processes must have the flexibility to deliver this kind of service and ideally make an emotional connection while doing so.

The flexibility to deliver a customized product is dependent upon a culture of trust. The company must trust and empower employees to read customers and deliver the product in the ways customers want and expect. To do that, employees need the right processes and technology.

A good example of this institutional flexibility is the new cellular provider Ting. This company has eliminated many of the standard unpleasant surprises customers encounter with their cell phones such as overage, late payment, and early termination fees. These surprises in cellular phone contracts, along with those in banking and credit cards, are described in a *Harvard Business Review* article entitled, "Companies and the Customers Who Hate Them."[1]

Ting provides painless product delivery, a positive emotional connection, and stellar service delivery in an industry riddled with long contractual commitments, service problems, and fierce competition. Competition is amplified by the fact that basic cellular service is so cheap that it is a commodity and the rapid innovation in smartphones between brands like Google, Apple, and Samsung have made even the handsets nearly identical. The only remaining differentiator is the CE.

Unlike its competitors who typically insist on a multiyear commitment, Ting offers cellular plans that scale up and down on a monthly basis depending on usage. If Ting customers use fewer minutes or megabytes than expected, they receive credits the following month. If a customer uses more minutes or megabytes than expected, they are not charged exorbitant penalties or overage fees; they are simply shifted up to the next tier of service. The product is automatically tailored to the customer's needs.

Ting's business model is a clear testament to the power and value of no unpleasant surprises. The company purchases satellite bandwidth from Sprint and packages it into a much stronger, friendlier service delivery. The result is happier customers and healthy profits, showing that service is especially important in commodity-like industries and that it can disrupt notoriously customer-unfriendly sectors such as cellular service.

Likewise, Zappos offers free returns, even to those who return a large percentage of their ordered shoes. Zappos is able to offer such favorable policies because of the creative and clever ways that it informs and educates its customers right at the point of sale, which all but ensures that returns are the exception.

Zappos carefully and preemptively addresses one of the most common customer questions ("Will this item fit me well?") with a Fit Survey containing aggregated information, answers to questions, and ideas from customers who purchased the item. This activity is now termed crowdsourcing. Shoppers are able to understand in specific but crowdsourced terms whether an item tends to run large or small for its given size. Zappos can offer such a favorable return policy specifically because it provides aggregated sizing information with its Fit Survey. The retailer does the job right the first time, painlessly informing and educating customers, thereby benefitting the bottom line by taking actions to minimize returns and to make the DIRFT phase flexible.

Encouraging Easy Access to Service via Effortless Channels

Easy access to service appears simple to many executives, but unfortunately it is not. Most companies look at access as a technological challenge: You answer the phone quickly, provide a phone tree to route the customer to the right group, and have a website that is responsive to email and chat requests.

As noted, customers often do not believe that companies want to help and therefore do not bother to make contact, and when customers do ask for help, they encounter multiple operational barriers. For example, one leading technology company received a significant number of calls around 2 A.M., when the techno nerds were working at full tilt but most of the company's employees were home sleeping. The company created an overnight support team to meet this need.

A company's approach to creating effective access should break down the barriers to customers seeking assistance and then ensure that the system is flexible enough to successfully handle their requests.

Breaking Down Barriers

Most customers do not request assistance because they do not know where to find help, they believe that the company will not help them, or they expect to encounter conflict. The access strategy must work to break down those barriers. Start with a message that is visible when and where customers are most likely to encounter problems—one that customers will notice. One brokerage firm printed, "We can only solve problems we know about!" on purple stickers placed on the front of the investment statement. In addition, the firm provided a clear, simple list of ways to receive service. Contacts from confused customers went up, which allowed the company to solve the problem and educate the customers.

In another example, a motorcycle company encouraged its dealers to put similar signs in their stores. When the company showed the dealers how much revenue they lost by not hearing about complaints, most dealers chose to post the signs.

Forecasting the Workload

The first challenge is to understand when customers will need help or want to contact the company. The primary rule is that people need help when using or thinking about the product. Contact workload depends on a wide range of factors—from volume of advertising and sales (turkey preparation at Thanksgiving) to the level of problems with the product and the need for customer self-assembly of toys on the night before Christmas. The weather and even the phase of the moon can affect calls too: More weird and emotional calls are received during the full moon (the word *lunatic* has an empirical basis!). Some financial service, utility, and many business-to-business (B2B) companies tend not to be open on the weekend, but questions are most likely to arise when customers are catching up on their mail and bills over the weekend. Your service system should be accessible when customers have their questions.

The next step is to make all desirable channels available exactly when they are needed. When the customer desires to self-service a product, the website map should clearly direct the customer to the proper location. When the customer wishes to change communication channels and talk to a human, that should also be clear and easy. For example, when customers are reviewing their printed or online credit card bills and find questionable charges, they want to talk to someone immediately. This means placing an 800 number on the printed statement or website or, better yet, a Call Me button on the website. Customers are happy to use self-service for tracking package deliveries until they encounter an unpleasant surprise (e.g., the package has not been delivered as promised). Customers want to simply click a button on the website and have a human call them with an explanation and a plan for recovery.

Creating the Capacity for Effective, Consistent Response

Accurate forecasting is the key to having adequate capacity across all communication channels: well-trained staff to handle calls; a website supported by an easily navigated, comprehensive, current knowledge base; and a phone tree that keeps up with changes in the organization to ensure timely, accurate answering of the phone, chat, website query, social media mentions, or email. However, even if executives were able to predict workload perfectly, they would need to create the effortless, secure channels to transport the customer to the well-organized, trained human staff equipped with the technology and manpower to handle that forecasted workload just as it arrives. This part of the process entails seamlessly traversing the chosen response channel to reach a service system that has the necessary capacity and response content.

Another major challenge nearly everyone now faces is complying with companies' ever increasing security requirements. Customers are more and more frustrated by having to remember (and change) their passwords despite their increasing concerns about hacking and identity theft. This creates more barriers to getting service, damages retention rates, and decreases positive word of mouth. To counteract this, many companies are now conspicuously designing their websites to provide answers to the most basic questions without requiring customers to log in and go behind the firewall. This is the smart approach.

Creating the Capacity to Respond to Changing Workloads

Staffing needs for answering the phone and providing an accurate response is a key challenge in a highly fluctuating environment. Any corporate call center and even your local dry cleaner face large fluctuations in workload. How do you cost-effectively have enough people to handle the Monday morning peak when, by 10:30 A.M., the workload has dropped by half and later dropped to one-fourth by Thursday afternoon? Customers will wait for service (access) only so long.

Technology has again come to the rescue for both the Password Hell[2] and fine-tuning online capacity. One successful solution was developed by Coursera, a blossoming online community that offers massive open online courses (MOOCs) from a variety of elite universities. Secure access is even more important in an academic setting because course credit is being awarded. Authenticating student work and verifying identities have been a challenge for both online learning programs and even traditional brick-and-mortar schools. An ineffective verification can place doubt on the accolades and credentials earned by the students, weakening the appeal of online courses for both aspiring learners and hiring managers seeking specialized skill sets.

Coursera's Signature Track provides a streamlined, effective way to verify online coursework without repeatedly asking for sensitive personal information, as many less informed industries tend to do. To add extra verification (and therefore extra validity, legitimacy, and credibility to the course), Signature Track requires a student to send in a scan of a government photo ID once at the beginning of the course. Coursera then uses a quick, painless verification process for all submissions and tests that involve a picture taken via webcam and the typing of a simple sentence to verify a typing pattern, determined by the speed and cadence of previous typing pattern verifications. This process is a valid verification method that would be difficult to circumvent but still eliminates the need to remember complex passwords. (Cheating this verification system would arguably be harder than simply studying and passing the course.)

A second access example is how nearly ubiquitous broadband has allowed contact centers to fine-tune their capacity to respond to radical changes in workload almost instantaneously via the use of home-based service representatives. Many call centers now have part-time home-based representatives who work less than a traditional eight-hour shift, such as 8:00–10:00 A.M. or 9:00–12:00 P.M. when the kids are at school. If a workload peak suddenly appears, the supervisor can notify 10 reps that there is work, and the first six who log on can then immediately start handling calls.

One flower delivery service created what it called microshifts of 45 minutes. While this works for large phone centers and online email and chat functions, it does not work for most retail service environments. Some retailers will make two- or three-hour shifts available for part-time staff, but this scheduling mechanism is harder for retailers than it is for companies whose employees can work from home.

Creating a Complete Service Experience
via Every Communication Channel

Fulfilling service requests is the traditional focus of customer service. Most service managers and frontline customer service representatives (CSRs) still see their role solely as responding to the individual customer and completing the transaction—for example, answering the phone, responding to email, taking and fulfilling the lunch order, or resolving the problem. In today's CE-oriented organization, the service delivery function should have five objectives:

1. First-contact resolution (the traditional objective)

2. Cross-selling when appropriate

3. Preventing future unnecessary service contacts via education

4. Creating emotional connection and value added when appropriate

5. Collecting information for the VOC

Very few companies are adept at achieving all these goals during a single transaction. Neiman Marcus Direct gets close because its staff members truly love the clothes they are selling and have the freedom and expertise to take the time to connect, educate, get feedback and log information, and cross-sell (e.g., accessorize) on almost every call. People are critical to great service, but so are processes, technology, and partners. Even if the service access function has made staff available, most companies do not have the process, technology, and culture for the employees to achieve all five of these goals. However, really good companies achieve all five in even the simplest transactions.

One example of an effective service delivery via technology comes from the service industry itself. Nosh, a social platform for frequent diners designed by Google-backed FireSpotter Labs, created a spinoff of Nosh called Noshlist, which is a simplified, streamlined restaurant wait-list process that does not involve large, cumbersome handheld buzzers with limited range.

Noshlist sends a text message to waiting diners' smartphones when their tables are ready. Diners are not required to stay within several hundred feet of the restaurant or hang onto a large electronic device. Restaurant owners do not have to purchase and maintain a separate electronic system. Beyond the basic and obvious advantages, Noshlist also offers extensive customer loyalty tracking, demographics, and analytics. If any guests waited a very long time to be seated during their last visit, hosts can acknowledge this past inconvenience, apologize, and seat the loyal guests more quickly or offer other gestures to thank them for returning after a less-than-ideal initial experience.

Additionally, Noshlist can tell hosts with great precision what their average wait times will be based on the time of day, day of the week, and history of service speed. This helps hosts manage expectations with accuracy rather than relying on guesswork, hunches, or friendly hospitality to compensate for longer than usual wait times and disappointed guests.

Listening and Learning by Creating an Effective VOC Process

Listening and Learning (L&L) is broader than the traditional VOC. L&L means receipt and analysis of all input received about and from the customers' experience. Traditionally, VOC has gathered information only from complaints, focus groups, and surveys. Now the input can include social media, input from employees who observe or participate in customer experiences, and even internal operations data that describes what your company process has done or is about to do to the customer. The learning aspect of L&L includes taking the VOC information and making sure the organization acts on what it heard. This is critical because, in my experience, more than half of all VOC processes have little impact because no one is paying attention or believes that there is a big payoff for taking action.

To effectively improve the customer experience, the L&L function must provide an end-to-end, unified picture of that experience. This process allows the organization to:

- Understand the CE of all the customers, including the large majority of customers who never contact the service system.

- Set priorities across all the organizational functional silos based on and propelled by an economic imperative for action.

- Improve the CE by assigning ownership of the implementation of the project to the function best qualified to lead the effort and then measure the effectiveness and financial impact of the recommended improvements.

The end-to-end requirement includes everything from the first presale marketing touch to the final billing, assuring that all unpleasant surprises are identified and that critical ones are addressed. For business-to-business relationships or ongoing services such as telecommunications or financial services, account executives or field sales representatives manage the ongoing transactions, account maintenance, and relationships. Often the problems not complained about do the most damage. For instance, in studies at three different copier companies, I found that being misled by sales reps causes four times the damage to loyalty as problems with machine breakdowns or delays in repair. However, the problem of misleading sales is seldom articulated as a complaint for fear of alienating the sales rep. Therefore, the most damaging problems are exactly the ones not highlighted by the complaint system.

A VOC process that provides a unified picture of the CE reduces the impact of gaps, contradictions, and biases in any single data source. Many companies define their VOC as solely customer surveys. To eliminate single-source bias, an effective VOC should include surveys, customer contacts (both service and sales), and relevant internal metrics describing what the company has done to, or for, the customer (such as warranty claims, missed appointments, out-of-stock reports, invoice adjustments, late charges, and missed shipment dates). Operations data can flag when a delivery will be late before the problem ever occurs. Customer calls, emails, and letters provide more actionable data because they are timelier than surveys and indicate, at the issue level, what the customer's expectations are and the problems most seriously affecting customer loyalty. Also, employee input is critical because it can explain the causes of problems and internal fire drills, in addition to describing incorrect customer expectations.

Two of the biggest recent changes in VOC are the advent of online communities and speech analytics, which are revolutionizing the VOC process. Online communities can give feedback and crowd-source priorities and ideas. For example, Starbucks and Intuit have reaped huge benefits by allowing their online customer community to answer many user questions and generate new ideas. Speech analytics, based on technology originally developed for the U.S. Intelligence Community, can interpret phrases and identify the purpose, subject, and even sentiment of a conversation. This can then be applied to thousands of calls daily to identify the phrases and answers that result in the highest consistent levels of satisfaction. The results can be used to train staff and evaluate their performance. One communications company has eliminated more than half its call monitoring staff, finding that speech analytics is more efficient and more accurate and consistent.

Applying Technology During
Each Phase of the CE Framework

The greatest single CE challenge is mass customization, that is, anticipating and tailoring the whole experience to the individual.[3] The only way this can be done, even in a small store environment, is to apply technology. Anticipating includes delivering what I have referred to earlier as psychic pizza. Anticipation is critical because it often is the opposite of an unpleasant surprise; it is a positive surprise that makes the transaction memorable as opposed to bland and impersonal. Further, even if it is not a huge surprise, it often eliminates uncertainty, which is a large source of dissatisfaction and unease. Think about the last time you were on a close airline connection and the captain announced that the airline would hold your connecting flight. What a relief! Customers want to feel in control. Proactive communication produces relief and delight in direct proportion to the importance of the event.

Tailoring to individuals requires knowing their history and personal circumstances. Again, technology is critical to both anticipating and tailoring the experience, especially given that most customers now want to interact by using it. If you bridle at this last suggestion, think about the last time you went into a bank branch to make a deposit or get cash from a teller; few customers want to take the time when an ATM will do the same thing in fewer than 60 seconds. When technology is easy to use, most of us want to use it for mundane tasks.

Psychic pizza can make even unpleasant tasks more pleasant. For example, we have all stayed home waiting for the cable technician or a repairperson. Suppose you have an appointment for the morning (8:00–12:00), and the company confirms the day before at 2:30 P.M., indicating that your technician will call between 8:00 and 8:15 with his estimated arrival time. Now you worry less and do not need to call the afternoon before and again that morning to confirm. If the technician actually calls at 8:15 and says he should be there at 11:00 A.M. plus or minus 20 minutes, you know you can run two errands without missing the appointment. Companies taking this approach have reduced calls from nervous customers who are anxious to get back to work, and they have achieved a higher percentage of fulfilled appointments due to fewer people dashing out for ten minutes to run an errand and missing the technician. Technology can facilitate scheduling and communication using the customer's preferred channel, which is determined at the time the appointment is first made.

Southern California Edison (SCE) provides another example of this. SCE emailed customers whose bills were going to be significantly higher (based on projections from the first ten days of electricity utilization data transmitted wirelessly) and suggested ways of reducing energy utilization. Over 50 percent of customers are opening the email (5 percent is considered a very good email open rate), and many have signed up for conservation support and reduced their energy use. Further, satisfaction with SCE's conservation actions has risen along with compliments—for sending them the bad news sooner.

Metrics to Measure and Manage Customer Experience Effectiveness

Most companies have only the haziest of ideas of how well they deliver on all four parts of the CE Framework, even though all four parts can contribute to higher satisfaction, greater loyalty, and a better bottom line. Executives often believe they have good metrics for each of these areas because they all consume significant resources. In fact, the area with the least overall potential for impact, service access, has the best metrics. The areas with the greatest overall potential impact, DIRFT and L&L, have the weakest metrics—or none at all.

Here are the performance metrics of most companies:

- For DIRFT, most companies have quality metrics for the products and services produced. These are the basic manufacturing and service operations quality metrics. However, few companies create metrics for how effectively customer expectations are set and how customers are educated on product use. If customers are misled or do not know how to get the most out of products, the value delivered is dramatically reduced. Think about your TV remote or all the features of your automobile computerized control center that you do not know how to use.

- Service executives have plenty of detailed metrics on service accessibility (abandoned calls and answer speed) but not on the effectiveness of complaint solicitation—that is, on breaking down the barriers to getting customers to tell you they are dissatisfied.

- The service delivery phase has the most robust metrics, usually based on satisfaction surveys and evaluations of recorded calls. In fact, service executives are drowning in a myriad of metrics for the actual service process—but many of the new ones are fads that may do more damage than good. For example, the Net Promoter Score (NPS), suggested by Fred Reichheld in the *Harvard Business Review*,[4] ignores the third of your market that contacts the company and is somewhat happy, as well as all the customers who have problems but do not complain.[5]

- Finally, very few companies have metrics for evaluating the effectiveness of L&L and VOC processes, such as asking what percentage of issues get fixed.

GETTING STARTED
Questions to Ask Yourself About Your Existing CE

1. Look at your existing CE process using the four-part framework. Do you know what your processes are for expectation setting, breaking down barriers, and the L&L?

2. Are you making the best use of processes to avoid predictable unpleasant surprises, or are you at least warning customers that surprises may or will happen?

3. Does your technology tell you where problems are going to occur?

4. Do you have metrics in place to tell you how all four of the CE components are working?

KEY TAKEAWAYS

There are four discrete components of a consistently great CE delivery mechanism:

- DIRFT (doing it right the first time) sets basic expectations for customers honestly and delivers the product or service as promised.

- Service access strategy breaks down the barriers to asking for service, keeps all possible communication channels available whenever customers need them, and makes them all easy to use.

- Service delivery makes an emotional connection with customers, prevents problems, and educates and gleans information while responding to the basic request for assistance.

- Listening and Learning uses a VOC process to measure the end-to-end CE and presents it as a unified picture that provokes action.

- Technology can enhance the performance of your people and processes in each of the CE components via anticipation, proactive communication, and tailoring the experience to each individual customer.

- You cannot manage what you do not measure and few companies measure all four CE components effectively.

Notes

1. Gail McGovern and Youngme Moon, "Companies and the Customers Who Hate Them," *Harvard Business Review*, June 2007.

2. Sam Greengard, "Escape from Password Hell," *CIO Insight Blog*, January 28, 2013.

3. Stan Davis, *Future Perfect* (Reading, MA: Addison Wesley, 1987), p. 150.

4. Fred Reichheld, "The Only Question You Need to Ask," *Harvard Business Review*, December 2003.

5. John Goodman, "The Passives Are Not Passive," *Quirks Market Research Review*, 25(10), October 2012.

CHAPTER 3

Jump-Starting Action by Quantifying the Revenue Cost of Inaction

Unless you make a convincing case, finance executives will starve your CE enhancement initiatives. Many are penny-wise and pound-foolish, failing to invest in CE because the revenue and profit payoff is not immediately obvious. As a result, companies often create processes that save money in the short term while alienating customers and driving away long-term revenue streams.

For example, if customers of one leading technology company call for assistance 91 days after purchase, and the customer does not have an extended warranty beyond the first 90 days, the company flatly refuses any support. This is because the finance and product management departments both believe that even one free support call would cost too much. My rough math suggested that the damage to loyalty of being rebuffed and the impact of WOM and the lost accompanying revenue were at least ten times the cost of the support call—but I argued to no avail. This company did not understand the math of future loss and potential long-term gain. Progressive companies help callers with the immediate problem and also inform the customer that an extended warranty must be purchased to get further support.

In another example, a doctor loves his smartphone and the fact that patients and other physicians can send him pictures of charts and medical conditions. When he was buying a new high-end auto, he tried syncing his smartphone with the communication system of his top choice. The car and phone would not sync, and the salesperson and the national service call center both blamed his phone. The next brand of car he liked instantaneously synced with the phone. Guess which car he bought? As his wife said, "Tom buys his phone because it is a good camera, and he selects his car because it is an effective phone." Most customers make decisions based on equally small details; companies that ignore minor problems can drive away huge amounts of revenue.

This chapter shows that a great CE has a top-line (revenue) impact far greater than the cost of delivering it. I also explain how to make the case to top management and create the economic imperative for immediately investing in service and CE. This is counterintuitive, but a great experience is almost always cheaper than a good experience. Finally, the chapter challenges the idea that you must always WOW the customer. In fact, WOW needs to be delivered only sporadically; always WOWing a customer is not cost-effective.

This chapter combines the implications of customer behaviors we examined in Chapter 1 (satisfaction, loyalty, WOM) with the concepts related to the four parts of the CE in Chapter 2. The chapter also shows how to quantify the bottom-line implications of the CE and investments in enhancing the CE so that the CFO will feel comfortable in making those investments.

We examine five topics:

1. Why do finance executives almost always believe that a great CE is more expensive than a good one?

2. How do you quantify the revenue, WOM, and margin impact of an enhanced CE?

3. How do you quantify the cost reduction impact of a great CE?

4. What are the mechanics of convincing skeptical CFOs and other functional executives not only of the value of a great CE but of the dangers of the status quo?

5. How do you pick your battles, that is, decide which of the myriad of actions to improve the CE make the most sense? Here we will also look at how technology can help you pick both strategic battles and make real-time cost-benefit decisions at the micro level.

Why Executives Believe a Great CE Is More Costly Than a Good CE

Unfortunately, short-term cost almost always trumps long-term payoff. I recently worked with two retailers and a home appliance manufacturer, who were all holding very low levels of inventory. Because the finance department assigned a significant cost to carrying inventory, the retail stores were short of sale merchandise, and the appliance manufacturer could not provide consumers with repair parts in a timely manner. The out-of-stock experience seriously damaged customer willingness to take the next sale seriously and to buy the same brand of appliance. In all three cases, there was a tangible cost of having the inventory, but the finance department could not see the *cost of not having the inventory*. A company can easily quantify the cost of providing service, inventory, and quality but often finds it difficult to quantify the revenue and WOM damage of *not* providing them.

A corollary is that a great CE is perceived to be more expensive than a good CE or than the current level of CE. This is due to a few standard outrageous stories used to illustrate a great CE. For example, a Nordstrom store once took back a set of tires when they did not sell tires. A delivery company staff person once hired a helicopter to deliver the package when the airport was fogged in. Although each of these events did happen once, they are not the usual great CE. Such expensive actions are neither generally required nor cost-effective.

Many people think quantifying the cost of a great CE is easy. They believe it includes staff, training, technology, and, especially, more refunds to unhappy customers. In fact, the reality is that customers are more satisfied with getting what they expected, and they feel treated fairly when given a good explanation of why it was not possible to give them what they wanted. This reality often requires counterintuitive actions. For example, one auto company found that when customers were given a full refund for an out-of-warranty repair, they were often not as happy as customers who had received free parts but had to pay for the labor to install them. This is because the partial payment for the repair parts had been accompanied by a detailed explanation of both the warranty and how the customer had already received full value from their warranty. The explanation was more important to satisfaction than the extra money.

Quantifying the revenue and WOM damage of a poor CE is not so easy for three reasons:

1. Revenue is in the future and viewed as speculative. The CE takes place right now, whereas the customer may or may not come back to spend more money sometime in the future. The finance department is appropriately cynical about future promises of revenue.

2. Both the finance and marketing departments believe that WOM is hard to measure and that its impact is even more speculative.

3. Costs of poor experience surface in other parts of an organization; for example, poorly set expectations surface as product returns and extra calls to the service department increase.

For these objections to be overcome, a methodology must be transparent, logical, and verifiable. The following approach meets all three of these criteria.

Measuring the Revenue Impact of a Great CE

The biggest challenge in quantifying the payoff of a great CE is quantifying its revenue impacts. This section first defines the three top-line impacts and then shows how to quantify the impacts so that even cynical CFOs will believe they are linked to the CE.

Components of the impacts to the top line include revenue, margin, and WOM. Revenue is the top line of the profit and loss statement. The top line impacts are calculated based on the value of the customer and the revenue, margin, and WOM impact of the CE. Each of these impacts can be measured and quantified.

Value of the Customer

Because all the revenue impacts are associated with winning or keeping customers, the prerequisite to quantifying the revenue impacts is to know what a customer is worth. If you do not know how much your customers are worth, you cannot calculate the value of winning or keeping them. When valuing the customer, most marketing and finance executives reference the Lifetime Value of the Customer (LVC), which was first mentioned in a 1986 White House study[1] but which Carl Sewell formalized as a term in his 1990 book, *Customers for Life*.[2] We both suggested valuing the customer as the discounted cash flow of all their purchases over 30 years of future purchases, by calculating either overall revenue or the net profit.

Since LVC's first introduction, many finance executives have expressed skepticism because, to quote one, "Who knows if we will even be in this line of business five years from now let alone in two decades?"

A better approach is to use a three- to five-year average revenue or gross profit flow as a conservative estimate of the value of a customer. If the customer buys only once every 10 years, then that revenue value would be half the value of a purchase. In many cases, it makes more sense to look at a household unit (all the members of a family living together in one house) rather than individual consumers because the experience of one will affect the behaviors of all the family members. Consider a household with two adults and one teenager who drives. Although any one person might buy a car every six years, together they will probably buy a car every other year, assuming that the purchases are evenly spread. Thus, in a five-year period, the household will buy two and a half cars, and the value of those purchases is the household's value to the company.

Impact of the CE on Revenue

Based on the hundreds of studies we have conducted in dozens of industries, a negative CE reduces customer loyalty by an average of 20 percent. This means that for every five customers with a negative experience, your company will probably lose one customer.

Some negative CEs, which I will also refer to as problems, will cause between 5 and 50 percent damage. The rule-of-thumb average for problem impact is a 20 percent decline in loyalty. For instance, at one snack food manufacturer, broken crackers resulted in a 5 percent decline in loyalty while stale crackers caused a 25 percent decline, and a foreign object or infestation (bugs in the box) caused well over a 50 percent reduction in willingness to buy that brand again. By combining this impact on loyalty with the revenue value of a customer, it is possible to quantify the revenue impact of a problem. For example, a problem causing a 20 percent impact on loyalty times a $100 value of a customer would produce a $20 revenue loss. You can then surmise that avoiding or preventing a similar customer problem would save $20. Eliminating or avoiding problems increases loyalty by an average of 20 percent.

Resolving an existing problem raises loyalty from 20 to 50 percent. The rationale of the range is that if the problem is serious but the company recovers effectively—by replacing the product quickly and assuring the customer that this issue will never happen again—almost all the loyalty can be recovered. In some cases, customers are even more loyal than before the problem occurred. However, in other situations, complete recovery is just not possible. The memory of bugs in the cracker box outweighs the benefit of replacement crackers.

Impact of the CE on Margin

Sensitivity to price doubles when a problem occurs, and then sensitivity doubles again if it recurs or multiple problems are encountered. For example, if 10 percent of customers who have not had a problem think the price is not a good value, 20 percent will think so after they encounter a problem. A significant percentage of those customers will not repurchase again when in the market for the product. Therefore, if a customer encounters a problem, high profit margins, which are dependent on premium prices, cannot be maintained.

On the other hand, if the customer is surprised and delighted, loyalty can be raised by as much as 30 percent. In addition, you can raise the price by 30 percent, and the loyal customer will keep buying. This is why Starbucks can charge high prices: Customers are delighted by the CE. However, if they encounter problems, they may decide their Starbucks lattes are not worth their cost.

Impact of CE on Word of Mouth

A negative experience usually causes two to four times as much negative WOM (or word of mouse on the Web) as a positive experience. WOM now entails much more than emails and postings on Facebook. Entries on review sites such as TripAdvisor, Amazon, and Expedia now have significant impact on many consumers' purchase decisions. Customer Care Measurement & Consulting's (CCMC) *2013 National Rage Study* found that 5 percent of the general consumer population post reviews; however, 30 percent of consumers rely on posted reviews to make purchase decisions.[3] CE management now must explicitly include fostering, measuring, and managing all aspects of WOM.[4]

Calculating the Top-Line Payoff of an Improved CE

A given CE will result in a certain level of customer loyalty; some revenue will be retained, and some will be lost when customers do not have perfect experiences. The loyalty level is a function of the customers who have a great experience, those who have a so-so experience, and those who have an unsatisfactory experience. The net revenue payoff of changing the CE must always be calculated as a change from the current or baseline situation. The baseline estimates the customers at risk from the baseline CE and how the enhancements reduce the revenue at risk.

The following discussion outlines the mechanics of first calculating the baseline level of revenue at risk and the result of moving from a current level of CE to an improved level. The mechanics include:

- Increasing revenue by retaining more customers.

- Winning more customers via WOM referrals.

- Increasing the profit margin from customers via a higher price based on more value and fewer problems.

Revenue Increase of an Enhanced CE

Perform the following steps to calculate the number of customers at risk due to the current CE and the revenue associated with the customer. After calculating the baseline revenue at risk, you may quantify the revenue impact of changing key parameters such as the percentage of customers who complain, who are satisfied, or who encounter problems.

Step 1: Quantify the revenue being left on the table due to the current situation

The Market Damage Model explained later in this chapter (see "Picking Your Battles") calculates revenue lost due to customers who are dissatisfied at the end of the CE. This calculation is based on all types of problems that customers encounter rather than on one specific problem. The Market at Risk calculation will show how to calculate the revenue damage of particular problems.

The Market Damage Model calculation makes the case for retaining customers by providing an overall picture of the impact of customer dissatisfaction. The model came out of the White House–sponsored study of complaint handling in both business and government agencies in the 1980s.[5] At the highest level, market damage is defined as:

- The amount of sales or revenue lost or at risk from customers who encountered problems, plus

- Sales lost due to factors other than problems (whether handled by the customer service function or not voiced at all). This damage is usually attributable to the lack of perceived value for the price paid.

In most companies, about half of all disloyalty is due to problems, and the other half is due to perceived poor value. So you must address both factors.

Figure 3-1 shows basic data on customers' problem experiences and resulting loyalty. You can see the full range of CE, the proportion of customers having each experience, and the loyalty and WOM impact of each possible outcome. The data is usually collected by asking a random sample of customers the following simple questions:

1. Have you encountered any problems?

2. If the answer is yes, the customer is presented with a relatively complete list of possible problems and asked to check all he/she has experienced.

3. Which problem was most serious?

4. Did you complain to the company about it?

5. If yes, the customer's satisfaction, loyalty, and WOM are recorded.

This short sequence of questions is all that is needed to build the chart in Figure 3-1 and the Market Damage Model.

Figure 3-1 shows the loyalty and WOM of customers who do not have a problem, as well as customers who have had problems and either complained or remained silent. The top line of the figure is sometimes perplexing because it shows that customers with no prob-

Figure 3-1

Snapshot of CE Highlighting Impact of Problem Experience

	I Question / Problem Experience	II Contact Behavior	III Contact Handling	IV Market Impact			
				% Very Satisfied with ABC	% Definitely Will Keep Purchasing	% Definitely Will Recommend	Word of Mouth
	No Question / Problem Experience 80%			81%	85%	69%	
Customers	Question / Problem Experience 20%	Contractors 25%	Satisfied 50%	82%	90%	74%	1.7
			Mollified 30%	52%	80%	52%	4.4
			Dissatisfied 20%	35%	70%	32%	5.5
		Non Contractors 75%		40%	75%	40%	2.9

lem are not always loyal and often do not tell anyone about their experience. The CE was adequate, but not everyone felt that they had received a great value for the price paid. Therefore, in this situation, 85 percent of customers will keep buying, but that implies 15 percent will probably buy a competitor's product next time. The solution to this issue is to add value to the product or to create an emotional connection, both of which will be addressed in Chapter 4.

This example is also indicative of many market situations in which customers have a problem but, if they are satisfied, they will be more loyal than if they had not had a problem. Beginning with a study I conducted for The Coca-Cola Company in 1978, we have consistently found that customers who are dissatisfied will tell more than twice as many people about an unsatisfactory experience than customers who have had a good experience.

Once the basic data on the customer problem experience and resulting loyalty is in hand (data collection discussed later), the revenue loss of the status quo of the overall CE can be calculated using the Market Damage Model (later in this chapter, I will show how to use the Market at Risk Model to estimate the damage of individual types of problems). This loss is the sum of two components. First, there is the revenue lost from customers who had no problem but will not definitely repurchase the product again because they do not consider it a good value. In this example, we have one million customers, and 80 percent did not have a problem; so 15 percent of those 800,000 customers would be lost due to poor value, or 150,000 customers.

The impact of the problems and service experience beyond the value deficit discussed above is calculated in Figure 3-2.

The Market Damage Model allows you to calculate the revenue payoff due to varying any of the three basic parameters:

- The percentage of customers with problems
- The rate at which customers complain
- The effectiveness of the service system in addressing reported problems

In this case, we assumed that 200,000 out of one million customers had a problem and that 25 percent of those customers com-

Figure 3-2

Quantification of Customer at Risk from Status Quo Situation

Getting the Resources: Quantify the Revenue Risk of the Status Quo

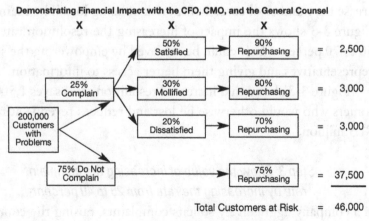

Demonstrating Financial Impact with the CFO, CMO, and the General Counsel

	X	X	X
		50% Satisfied → 90% Repurchasing	= 2,500
200,000 Customers with Problems	25% Complain	30% Mollified → 80% Repurchasing	= 3,000
		20% Dissatisfied → 70% Repurchasing	= 3,000
	75% Do Not Complain	→ 75% Repurchasing	= 37,500

Total Customers at Risk   =   46,000

at $1,000 per Customer, $46,000,000 at Risk

plained, 50 percent were completely satisfied, and 90 percent of completely satisfied customers will repurchase and are retained. Mollified customers are less loyal, and complainants who are left dissatisfied are the most disloyal.[6] We see that 46,000 customers are at risk of being lost. Noncomplainants make up the vast majority of this total, and this is the case in almost every organization. The largest opportunity to retain customers is to either prevent or surface the problems of those who currently suffer silently.

Once the revenue implications of the status quo have been quantified, you can start addressing the payoff of strategies to improve the CE and reduce the number of customers at risk. Three strategies are implied by the model:

- Better problem handling will increase the satisfaction and loyalty of customers who complain.

- Greater service system accessibility will get more customers to complain.

- Problem prevention will eliminate the cause of dissatisfaction.

Step 2: Show the value of increased effectiveness of problem handling

The model consistently demonstrates that increasing the percentage of customers who are completely satisfied after complaining will decrease the number of customers at risk of being lost. The example in Figure 3-3 shows the impact of increasing the resolution rate from 50 to 70 percent. This could be achieved by empowering the service representatives and giving them better access to information.

Figure 3-3 shows that increased resolution rates saves 1,500 customers who would otherwise be lost and reduces revenue at risk by $1.5 million.

Step 3: Show the value of increasing the complaint rate by increasing the rate from 25 to 40 percent

If a company aggressively solicits complaints, raising the complaint rate to 40 percent of their customers, or 80,000, phone calls increase

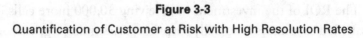

Figure 3-3

Quantification of Customer at Risk with High Resolution Rates

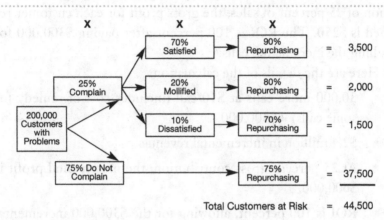

to 30,000. Operations and finance executives often view this as un-necessary work. However, if the company satisfies the customer, it gains significantly more revenue than it spends on the service trans-action. Figure 3-4 assumes a service cost per incremental complaint handled of $10.

Figure 3-4

Quantification of Customer at Risk with Higher Complaint Rate

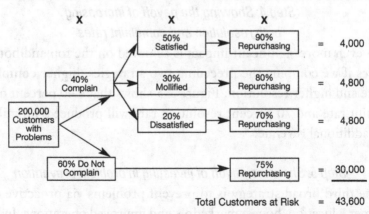

at $1,000 per Customer, $43,600,000 at Risk or $2,400,000 Savings
at Gross Margin of 25%, 30,000 Additional $10 Calls Give ROI of 100%

The ROI of the investment in receiving 30,000 more calls from dissatisfied customers is actually surprising. Assuming a gross contribution of 25 percent of sales, the gross profit for each customer retained is $250. The ROI is 100 percent after paying $300,000 for handling the incremental calls.

Here are the details of the calculation:

- 30,000 more calls at $10/call (includes cost of remedy for some calls) = $300,000

- $2.4 million in incremental revenue

- At 25 percent gross contribution, the incremental profit is $600,000.

- ROI is 100 percent, allowing for the $300,000 incremental cost of handling the calls.

The marketing, finance, and operations departments will often all say, "Why get extra calls?" This analysis shows that incremental calls from unhappy customers who would otherwise leave provide a positive return on investment. Another response is to ask the marketing department what the cost of winning a new customer is. If the cost is greater than $20 per customer, it is more cost-effective to keep a customer via service than to find and win a new one.

Step 4: Showing the payoff of increasing both resolution and complaint rates

An even more significant impact is achieved on the top and bottom lines if we combine the preceding two strategies, higher complaint rate and higher resolution. Figure 3-5 shows that a 40 percent complaint rate and 70 percent resolution rate will produce $4.8 million in additional revenue.

Step 5: Show the payoff of investing in problem prevention

The third broad strategy is to prevent problems via proactive customer education, honest marketing, and improved operations. In Figure 3-6, the reduction of problems by 25 percent produces a huge

Figure 3-5

Quantification of Customers at Risk with

Higher Complaint and Resolution Rates

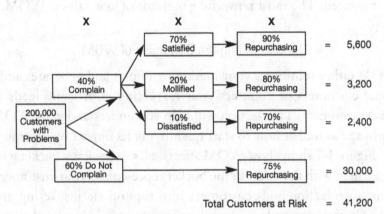

at $1,000 per Customer, $41,200,000 at Risk Adding $4.8MM to Top Line
Net Enhanced Top Line of $4.8MM in Revenue and $1.2MM Profit,
Which More Than Offsets 30,000 More Calls at Cost of $300,000

Figure 3-6

Quantification of Customer at Risk with Problems Prevented

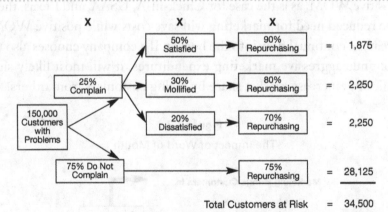

at $1,000 per Customer, $34,500,000 at Risk or Saving of $11,500,000
at Gross Margin of 25%, $1MM Spent on Prevention Gives ROI of 187%

reduction in the number of customers at risk. This shows that spending $1 million on prevention activities such as customer education produces a large return on investment.

This strategy provides the greatest impact and is usually the most cost-effective.

All of these strategies are based totally on sales revenue that can be measured. The most powerful generator of new sales is WOM.

Understanding the Dynamics of WOM

WOM either reinforces your marketing efforts and generates additional customers or counters your marketing efforts and leads to fewer customers. This section will help you understand your WOM landscape as well as how to start quantifying its impact on revenue.

Figure 3-7 shows how WOM affects sales. It depicts a bucket with a hole in it, with the water in the bucket representing your customers. Your organization adds customers into the top via marketing and sales. Customers who have a good experience remain loyal and attract more customers through WOM. Customers who encounter problems and are not satisfied leak out through the hole in the bottom of the bucket. Problem prevention and tactical customer service reduce the size of that hole. If a company experiences very little leakage and positive WOM, as is the case for Chick-fil-A, USAA, and Lexus, then the reduced need for marketing will save costs while positive WOM creates a continual flow into the bucket. If a company chooses also to continue aggressive marketing expenditures, it will most likely significantly increase market share by taking customers from others.

Figure 3-7

The Impact of Word of Mouth

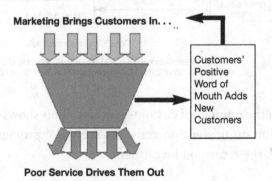

WOM has been called the most powerful form of marketing. If a report of a CE goes viral, it can cause significant benefit or damage depending on whether the report is positive or negative. However, viral events are the exception, and some mitigating factors suggest that the finance department is right to be conservative in its consideration of investments in social media programs.[7] Six dimensions of WOM affect its actual impact on revenue. Figure 3-8 helps you understand them:

1. *Channels used*—Virtually everyone interacts face to face and almost everyone interacts via telephone. The technological channels are very broad, with 61 percent of consumers using smartphones with Web access.[8] This penetration varies significantly by age with 78 percent of twenty-somethings and 42 percent of those over 55 having smartphones.

2. *Use of channel for complaining*—CCMC's *2011 National Rage Study* shows that most consumers prefer to complain directly to the company using traditional channels, with only 11 percent going to a social channel even for their most serious problem.[9] *The 2013 National Rage Study* found that only 8 percent of consumers post on a review site, while 30 percent read online reviews before a major purchase.[10]

3. *Reach of channel*—My research at CCMC and at TARP found that when consumers complained via face to face or phone, they seldom told over 20 other people. On the other hand, the *2011 National Rage Study* found that the average person with a social media site such as Facebook had 280 friends. This varied from an average of 334 friends for twenty-somethings to 53 for those over 65. The conclusion is that few customers immediately go to a social media site with a complaint, but those who complain on social media can tell a lot of potential customers about the problem.

4. *Magnitude*—Recent studies have continued to support the traditional findings, first published in the study for The Coca-Cola Company in 1978 mentioned earlier. Again, this study

Figure 3-8

The Landscape of Word of Mouth

Type of WOM	Channels Used	Who Uses	Reach of Channel	Use for WOM
Personal WOM	Face to Face	Everyone	Circle of Friends	Everyone
	Telephone	95%+	Circle of Friends	Everyone
Online WOMouse	Email	70%	Select Friends	Most
	Personal Blogs	5%	100	Everyone
	Social Network Sites	25%	230	25%
	Post on Review Sites	8%	Potentially Millions	10%
	Read Review Sites	30%	Potentially Millions	Potentially Millions

found that people tell twice as many others about a negative experience than they do about a positive experience. Surprisingly, this ratio generally holds even in large business-to-business relationships.

5. *Extent of WOM*—We have found that the descriptive details that consumers provide other consumers have a great impact on the utility, credibility, and therefore the impact of the WOM. Think of your own reaction to detailed stories at a cocktail party, as well as which reviews on Amazon you find most useful. Intelligent detail is important.

6. *Action taken based on WOM*—You can obtain an estimate of the impact of WOM by asking customers, "To your knowledge, of the people you told, how many of those told actually took action on your recommendation?" One-third to one-half of respondents, both consumers and those in a B2B environment, indicate that they are able to estimate the number and percentage of those receiving their WOM referrals that took action.

Quantifying WOM's Impact

Quantification of the impact of WOM is still a relatively speculative art. Part of this is due to understanding the motivation of consumer purchases. For instance, an executive in the movie industry rolls his eyes when WOM is mentioned. He relates a consumer research study his company conducted in which 20 percent of consumers reported that they had attended a movie due to the advertising of the film on television, but *there were no television ads*. Thus, consumer motivational research is an inexact science.

WOM can impact sales in two ways. First, if consumers in the market for a product get a direct recommendation for or against a particular brand, there is a high probability that the recommendation will affect their decision. Second, if customers specifically intend to buy a category of product, they might go to a review or social media site to learn about the product. The CCMC *2011 National Rage Study* found that 36 percent of women and 21 percent of men said that they would go to a site before making a major purchase. Interestingly, only 8 percent of consumers said they would post feedback on a review site.

The two approaches to the calculation would be either (1) to quantify the total volume of WOM produced by the CE and then multiply by the estimated impact or (2) to calculate the WOM volumes separately for positive and negative experiences and then estimate the aggregate impact.

Figure 3-9 shows the overall impact of servicing 10,000 customers in a week of interactions. When 10 percent are delighted, 80 percent satisfied, and 10 percent left dissatisfied, it is possible that the net WOM will be positive. If negative WOM has twice the impact of positive WOM, it is possible that the 333 newly won customers could be more than offset by the 6,000 negative reviews.

The second more detailed approach is to calculate the impact of WOM from a particular customer in the following manner. First, subtract those who are not in the market for the product at that time from the number of customers who hear the WOM. For instance, if you are discussing golf clubs and only 10 percent of people play golf, the WOM is irrelevant to 90 percent of those hearing about the CE. Second, estimate how many of those who are in the market actually take action on the WOM.

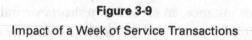

Figure 3-9

Impact of a Week of Service Transactions

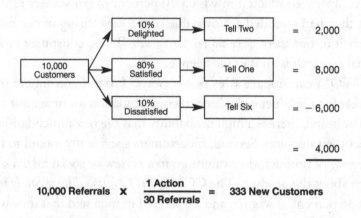

In several recent studies, when customers were asked to estimate how many people acted on their recommendations, only 33 to 50 percent of the survey respondents indicated that they felt comfortable estimating the number who took action. When this happens, you must decide how to calculate the impact of WOM. If only one-third of the customers spreading WOM felt comfortable estimating how many people acted on their referrals, the conservative approach is to attribute impact to only one-third of the customers. This approach will appear conservative and win finance department support while still allowing you to count some WOM impact. Figure 3-10 illustrates this approach to the calculation, in which we attribute market actions only to the customers who were positively identified by survey respondents as having taken action. The alternative approach is to extrapolate the WOM impact to the total number of customers told by all customers.

A similar calculation must be executed for negative CEs to estimate the WOM impact from the CE on net sales. This second approach is a more conservative method because it explicitly addresses the relevance of the WOM to the customers who are hearing about the CE. This method can also address the fact that some customers surveyed may have no idea how many of those they told took any action.

Figure 3-10

Calculating the Impact of WOM

100 consumers told about the positive experience	x	10% of consumers who were in the market for product	x	40% of consumers were estimated to have acted on what they were told	=	100 cases of positive WOM result in 4 new purchases

Both of these approaches are somewhat speculative, but technology is now allowing the testing and verification of the underlying premises. There are companies and tools, such as Google Analytics and Resonate, that can track who hears what online and then verify via large panel studies how many customers actually act on what they heard. In many cases, the modeling can even help you understand the causality behind the action. This is a technologically turbocharged replication of traditional social science quasi-controlled experimental design. In this approach, three messages would be sent to three samples of 10,000 customers to see which had the biggest impact compared to a control group. With Customer Relationship Management (CRM) software, you can track purchases and link back to most if not all the touches the customer has received.

Quantifying the Impact of CE on Margin

Most customers will pay more for higher quality. In fact, my research shows that in most markets, including the retail banking example in Figure 3-11, sensitivity to price is strongly correlated with problems.

Figure 3-11

Problems Raise Sensitivity to Price

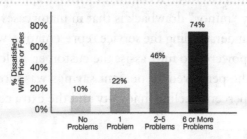

Fewer problems result in lower sensitivity to price, as shown by this survey of more than 3,000 retail banking customers.

This means that companies with better service can achieve higher margins. Customers may say, "You're expensive, but you're worth it because I seldom have problems." Margin is very important to the finance department because small changes in margin can create a large impact on profits. For example, if the average profit margin of U.S. companies is 8 percent, then a 1 percent increase in margin will produce a 12.5 percent increase in corporate profits.

Additional Revenue from Sales Due to Recovery

The final impact on revenue comes directly from the problem recovery process. In many cases, when the company says it is sorry, it will compensate the customer with a coupon for future free services or a discount on future purchases. In almost every case, the customer ends up spending more than the coupon is worth, producing additional revenue for the company. For instance, at Amtrak, for each $100 "I'm Sorry" coupon for free travel that was given out, the consumer spent $300 more toward future travel. So the net result was a gain of $200. My joke to the CEO was that if they could give out enough coupons, they could double revenue. At one hotel chain, the CFO calculated that every dollar given in coupons produced $7 in additional profits.

Quantifying the Savings of a Great CE

Most executives view the cost of a great CE to be primarily staff salary and refunds to customers. This is the genesis of the movement to outsource call centers to foreign countries with cheaper labor at half the cost. The only "minor" drawback is that in many cases the customer has difficulty understanding the service representative, who is not well trained or empowered to fully assist the customer. At least six studies showed that the perceived 30 percent savings was more than offset by the 20–40 percent decline in loyalty and the extra cost in terms of the extra work due to escalations to supervisors and headquarters, higher legal and regulatory costs, and higher turnover among good but frustrated local employees.

The aspects of a great CE that can lead to a measurable impact on organizational cost and therefore the bottom line are described here, along with the approach to quantifying each for the finance department. Not all aspects will be practical in all settings.

Reduced Service Workload

In most companies, 30 percent of service contact disputes could have been prevented by better communication up front. In one insurance company, the top ten reasons for calls from client companies were about account maintenance. Six of the ten reasons were due to poor client understanding of the account maintenance process. The insurance company service department then proactively educated 100 corporate clients about their account maintenance process when they called with a question. Further, they emailed so-called tip sheets with guidance on how to avoid the most common account problems. When we compared the number of calls from the 100 educated clients to the comparable call workload of a control group of 100 similar companies, we found that the number of calls from the educated clients had in fact been reduced by almost exactly 30 percent.

To quantify the preventable workload, you must identify the percentage of service transactions that could be avoided if customers fully understood the product and used it correctly. Then quantify the cost of that workload to justify investment in proactive customer education as in the previous case study.

In the case of the insurance clients, education meant creating the tip sheet and then investing three minutes of phone time to conduct initial education and introduce the tip sheet. The payoff was a 30 percent reduction in calls on a range of topics. Also, the company research indicated a significant increase in satisfaction when clients called less. A small investment in education created higher satisfaction and lower workload.

More Efficient Service Transactions
Due to Better Tools and Explanations

Examine and quantify the type, volume, and cost of the calls that are escalated or passed to a supervisor, as well as repeat calls from the same customer on the same day. You will find that many transactions requiring multiple calls or escalations are caused by obvious process barriers to resolution. For example, if the auto body shop says the repair will take five days, the customer can become angry. If the manager explains that it takes one day to get the parts and that three days are needed to apply four coats of paint and a 12-hour drying period between coats and that another day is taken up to buff the surface, the customers are not happy, but they are much more willing to accept the explanation because they understand the reason. Likewise, immediate access to a database to provide an answer is always dramatically more efficient than saying that you need to request an answer from another unit (or approval from a supervisor) and call the customer back. As you eliminate the barriers to the customer service representative resolving or mitigating the problem, not only do you improve the CE, you improve representative efficiency.

Lower Warranty Costs by Educating
Customers on Product Use and Troubleshooting

One electronics company found that over 70 percent of products returned for repair had nothing wrong with the product. Once the troubleshooting section of the manual was expanded (both printed and online), the returns decreased significantly.

Identify the top three customer-caused returns or problems and quantify how much is spent on warranty costs to address them. Then determine the cost of proactively educating customers on how to avoid each of those three problems. For example, posting cautions about the three top preventable issues on the homepage of the website will result in a reduction of those issues by at least a third. Although you will not prevent all occurrences of the problems, you should be able to cut their occurence by at least a third.

Lower the Problem "Remedy" Expense with Better Communication

A major faucet company has found that offering options for receiving parts that the customer can install or a discount on a new (often more expensive) product can often be more effective than giving the customer a full refund or reimbursement. Because even faucets are rapidly becoming more sophisticated, containing movement and heat detection technology that allows activation without turning a valve, customers can be easily convinced that they have received fair value and that getting a discount on a new, more sophisticated faucet often resulted in higher satisfaction than just immediately paying for a replacement. For this strategy to work, the communication of the options and warranty must be clear and not misleading and the incremental value of the new product must be clear.

Lower Marketing Costs

Ask the marketing department to estimate the number of new customers they win versus the number of customers derived from WOM referrals. Second, estimate the average cost of winning those customers using traditional methods such as advertising and promotions. This analysis will usually show that fostering WOM with great service is five times more cost-effective than further investments in traditional marketing. Using the estimates of enhanced WOM, estimate the degree to which WOM could lower the costs of the marketing department. Avoid the internal political problems this analysis might cause by suggesting that WOM can help enhance market share rather than decrease the marketing budget.

Reduced Costly Legal, Regulatory, and PR Disasters

These costly situations almost always start as unresolved complaints that get escalated to lawyers, regulators, and the media. If the customers' issues are not resolved, a small percentage of dissatisfied customers, usually 2–6 percent, will complain to a regulator, lawyer, or a member of the media. Therefore, if companies can find and resolve problems before customers contact lawyers or regulators, these expensive occurrences can be minimized. An effective L&L process can leverage the results of the VOC to reduce these contacts significantly by highlighting issues and then preventing them from occurring—or at least mitigating their impact before a complaint is made.

For example, in one health care system, the institution of a more aggressive complaint handling system that surfaced and quickly resolved complaints led to a reduction of insurance and lawsuit payouts of more than 10 percent. The legal department of one medical device company stated that they viewed the $5 million annual expenditure on the aggressive customer education and complaint processes as an insurance premium protecting the company from receipt of an FDA warning letter. In the past, the receipt of such a letter had required compliance expenditures approaching $100 million.

Lower Turnover Among Frontline Staff

Employee research for financial services and retail companies has shown that up to 50 percent of voluntary turnover among good frontline employees is due to frustration in handling preventable customer problems. Talk to your human resources department about the costs of turnover and quantifying the degree to which voluntary turnover is due to such frustrations.

Convincing the CFO

The finance department and other skeptics usually bring a series of objections to these calculations. Common objections and effective responses are:

- "Most customers complain." This is flat-out wrong. Major corporate customers in the chemical, shipping, aerospace, and large commercial finance industries have all indicated a failure to complain because they believed complaining would do no good or feared poisoning the atmosphere. Instead of complaining, they often cut back in discretionary spending.

- "All our good customers complain." We find that, although top customers will complain a bit more than the average customer, they often do not complain for the reasons described and because they expect to be doing business with the offending company for at least a year or longer.

- "It is too expensive to keep customers, and it is cheaper to win new customers." This is where you challenge the marketing department to quantify the cost of winning a new customer and then compare it to your cost of recovery. The Market Damage Model will demolish this argument.

- "Customers do not follow through on threats to switch brands." CRM systems and loyalty programs now provide the ability to track exactly what customers purchase. Based on a half dozen long-term loyalty tracking studies we have executed, at least 50 percent and usually more than two-thirds of customers do follow through on their threats to switch brands or suppliers.

- "Some customers have little choice; they are stuck with us." Customers of monopolies, nonprofits, government agencies, and other companies with long-term contracts often find ways to either switch brands or punish the organization that has provided the poor CE. How does this model work in their environment? In general, such customers will take three actions, all of which hurt the organization and create extra cost:

  - Complain to executives or outsiders such as legislators or the press

  - Stop using the product or service and cease to buy or use other services from the same company

  - Argue about charges or, when money is tight, place your organization lower on the list of who gets paid

All three of these behaviors will do damage to your organization in terms of both short-term cost as well as the long-term volume of customers. Also, as soon as an alternative does become available, unhappy customers will leave in droves.

Picking Your Battles

The most challenging problem you will face once you get your fellow executives on board to support a great CE is picking your battles. Fortunately, the Market at Risk methodology, developed during my research with Xerox and Shell Canada in the 1980s, will allow you to select the opportunities with the biggest payoff in a manner that the finance department will understand and accept. This model is the follow-on model to the Market Damage Model and was developed out of work for Xerox and Motorola in the late 1980s. It still applies and works well today.

If you have already built the Market Damage Model discussed in the previous section on quantifying the revenue impact of a great CE, you already know the overall frequency of problems, the frequency of occurrence of specific types of problems, and the damage that each type of problem does to customer loyalty. You can then calculate the percentage of customers who are at risk for each type of problem using the formula in Figure 3-12. This is the Market at Risk calculation.

Figure 3-12

Market at Risk Calculation

Overall % customers experiencing a problem	x	% customers experiencing a particular problem	x	% customers not likely to buy product again	=	% customers at risk due to that particular problem

An example of Market at Risk Model results is presented in Figure 3-13. The chart is based on slightly modified data from a technology company that won the Malcomb Baldrige Quality Award more than a decade ago.

Figure 3-13 shows that the most prevalent problem is meeting promised delivery dates and that 10.5 percent of the customers who encountered this problem are at risk for a total of 1.3 percent of the total customer base. If the customer base was worth $100 million, you

Figure 3-13

Market at Risk Model Results

Overall % w/ Problem (45%)	% Problems	% Won't Recommend	% Customers Potentially Lost
Meeting promised delivery dates	27	10.5	1.3
Product availability within desired time frame	23	0.0	0.0
Meeting commitments/ follow-through	21	30.0	2.8
Equipment/system fixed right first time	20	22.2	2.0
Adequate post-sale communications	19	10.0	0.9
Returning calls	16	33.3	2.4
Minimum customers at risk			**9.4%**

could say that $1.3 million is at risk due to this problem. Where the analysis gets very interesting is the second most prevalent problem: product availability. Here we see that even though product availability is a prevalent problem, this issue is causing no market damage; customers are willing to submit their orders more in advance with no damage to loyalty. This is a squeaky wheel that deserves no grease. On the other hand, the last issue on the list, sales reps not returning phone calls, does serious damage. The company can set priorities based on the payoff of reducing or eliminating the problem.

GETTING STARTED

Questions to Ask Yourself About Your Business Case for CE

1.   What is the average value of the customer that the finance department would accept?

2.   What is the cost of winning a new customer?

3.   What is a conservative estimate of the size of the noncomplaint rate? If you have no basis for estimation, have your service representatives ask 100 customers if they have had problems they have

not articulated to your company and use that percentage as a starting point.

4. Does the marketing department agree that WOM is important?

5. What are the top three preventable problems and the cost of their remedies? Extrapolate to the market based on the noncomplaint rate estimated in step 3 and the revenue based on the value of the customer identified in step 1.

6. Are there good examples of customers lost due to unarticulated problems? Personally call ten customers who have recently left your business and ask them why they left. You will identify at least five customers who had problems but did not complain, and these stories will be powerful ammunition, especially if you record several of the conversations and play them back for management.

7. Are there examples of significant warranty or service costs incurred due to preventable customer problems?

KEY TAKEAWAYS

• Finance executives do not invest in CE because the revenue and WOM payoff are hard to quantify, whereas the required up-front investment is easy to estimate. The top-line revenue benefits of a great CE include retained customers, positive WOM, reduced negative WOM, and higher margins. The ROIs of investments in an enhanced CE are usually over 100 percent.

• A simple rule of thumb for estimating the revenue impact of a particular issue is to assume that when a customer encounters a problem, on average, loyalty is diminished by 20 percent. Therefore, for each five customers who have problems, one will be lost the next time they have a purchase opportunity.

• Cost savings include reduced service workload, more efficient service transactions, lower problem remedy costs, lower warranty costs, reduced regulatory interventions, reduced litigation and PR costs, and lower HR costs due to turnover of good employees.

- Even after you have quantified the revenue impacts and savings of CE, the finance department and other skeptics usually bring a series of objections to the data. Be prepared to convince them by using the arguments in this chapter.

- Not all problems are created equal. Usually two or three problems do the most damage to loyalty, and some squeaky wheels do not deserve grease. Pick your battles.

Notes

1. John Goodman, Marc Grainer, and Arlene Malach, *Complaint Handling in America* (Washington, DC: White House Office of Consumer Affairs, TARP, 1986).

2. Carl Sewell, *Customers for Life* (New York: Random House, 1990).

3. Scott Broetzmann, Marc Grainer, and John Goodman, *Results of the 2013 National Rage Study* (Alexandria, VA: Customer Care Measurement & Consulting, 2013).

4. John Goodman, "Treat Your Customers as Your Prime Media Rep," *BrandWeek*, September 12, 2005, p. 16.

5. Goodman, Grainer, and Malach.

6. Completely satisfied are those who give a "top box" rating on a satisfaction survey. Mollified are those giving the second and third (middle) box rating on a five-point scale of satisfaction.

7. John Goodman, "Don't Go Overboard on Social Media," *Marketing News*, October 2011.

8. Todd Wasserman, reporting on new Neilson study, *Mashable.com Blog*, June 6, 2013.

9. Scott Broetzmann and Marc Grainer, *Results of the 2011 National Rage Study* (Alexandria, VA: Customer Care Measurement & Consulting, 2011).

10. Broetzmann, Grainer, and Goodman, 2013.

2

Designing the End-to-End Customer Experience

Do It Right the First Time (DIRFT)

The key to delivering perfection is to clearly promise what you intend to deliver and then deliver exactly what you promised with flashes of brilliance. To achieve that perfection, you must do three things. First, set customer expectations honestly. Second, design a system that is flexible enough to consistently deliver on the promises as modified by the diversity of individual customer needs. Third, create a culture that fosters inexpensive flashes of emotional connection. The key to the success of all three—consistency, flexibility, and a culture of connection—lies in your technology.

Technology can facilitate the achievement of all three aspects of success. For example, we have all waited at home for repair technicians. Most customers make repetitive confirmation and status calls the day before and day of the appointment to make sure the technician is really coming. New Jersey Natural Gas (NJNG) saves time, expense, and customer frustration by identifying the customer's preferred channel of communication when the appointment is made; by automatically sending confirmation via that channel the day before; and by providing an updated estimated time of arrival on the morning of the appointment. The NJNG approach is not only a WOW, it makes sense. The customer does not waste time waiting and wondering, and NJNG saves the cost and resource requirements of fielding several phone calls.

Technology can also actually deliver service flexibly, without unpleasant surprises. For example, when my wife and I go to the movies, we now buy tickets online, select seats upon booking, pay for parking through a mobile app, and pick up the tickets at the kiosk. Only in the theater do we interact with an employee, when a quick, friendly "Enjoy the show!" with eye contact is a nice touch. We can even order refreshments at an electronic kiosk, though a person still hands them to us. Technology does not create value beyond preventing unpleasant surprises (such as getting stuck in the first row)—but that's all it needs to do.

In this chapter, you will learn how to:

- Execute the essential steps of DIRFT without building in unpleasant surprises.

- Build a successful, customer-focused culture achieving both DIRFT and remarkable connection.

- Deliver on the promise every time with flexible processes.

- Establish the right metrics for managing and evaluating DIRFT.

DIRFT: The Essential Steps

The basic goal of DIRFT is to provide an end-to-end customer experience that delivers value without any unpleasant surprises. Although many companies deliver the better mousetrap, they often simultaneously negate it with dumb glitches in other parts of the experience.

Reviewing how your organization performs all the activities, end to end from product design to ongoing product use, will ensure that unpleasant surprises have not been built into the CE. A well-designed DIRFT delivers remarkable value in at least some activities and without unpleasant surprises in any. Books have been written on each activity, but their impact on CE has rarely been addressed. Here I will define the activity, highlight the most common unpleasant surprises to avoid, and give examples of innovative approaches that deliver remarkable value—some by using technology and some by the simple actions of people.

The best approach to reviewing DIRFT activities is to create a process map that shows, from the customer's perspective, how the customer hears about, buys, takes delivery, and uses your product. In this map, you then build in all the touches the customer has with your organization and any partners such as agents or delivery vendors. Executives of your company are not qualified to create this CE process map because they do not know all the details of the processes your company imposes on the customer. Your best consultants for this process mapping exercise are the frontline staff from each functional department and your Lean Six Sigma or quality staff.

If you review your company's performance in each activity from a CE perspective, I guarantee that you will find that at least two unpleasant surprises have been built into your CE. Finally, this map should provide the basis of the discussion, suggested later in Chapter 8, that you have with your information technology (IT) department on how best to apply technology to all parts of the CE.

Designing a Product That Improves on the CE

Most improvements address drawbacks in current products. For example, at Avis many years ago, the biggest single problem for customers was turning the car in at the airport and missing the plane because of the wait in the rental return check-in line. Once the damage to revenue was quantified, Avis thought out of the box and started using wireless handheld computers to check customers in, eliminating the lines and accompanying delays. This approach is now ubiquitous in the car rental industry.

Two leading heavy industrial companies, one a chemical shipping terminal and the other a paper manufacturer, wanted to differentiate themselves from the competition. In both cases, the industry norm was to ship the product by third-party carriers on trucks or train cars, tell the customer when it was shipped and by what carrier, and then let customers discover and deal with delays. When the companies began tracking shipments and notifying customers of delays, customers loved it. One customer especially appreciated getting an email if the shipment was going to be up to two hours late and a cell phone call if it was going to be later. In each company, a staff person now spends an hour each evening logging in to the carrier tracking systems from home to check the progress of shipments. Although this is a little more work than the old system was, the process change is a significant competitive differentiator.

A standard problem in product design is making the product so complicated that the customer is overwhelmed and has difficulty doing simple tasks. The TV remote and your auto digital control center are two examples. Elegance is often achieved by deciding what features not to include. The VOC process should be your primary guide to where improvements are needed, as well as for ideas for new products or functions. A second mistake is building the product so cheaply that it seems to be of lower quality. For example, airplanes, major home appliances, and faucets are all made with lots of plastic and reinforced carbon fiber, and many consumers view plastic as less durable than metal. Unless you communicate the rationale and improved value, such as strength and durability, the change is an unpleasant surprise. The key to success in design is to enhance value without any accompanying unpleasant surprises.

Creating Awareness of the Product and Its Value

Creating product awareness has the objective of putting the product message in front of customers in a way that leads to purchase consideration. It usually entails traditional advertising and marketing or fostering a critical mass of WOM. The most common negative customer experiences with advertising and marketing are invasive messages: telemarketing calls, loud TV ads, poorly targeted messages (such as email spam), and initially misleading messages.

The issue of spam is often one of degrees. I am the member of several associations that significantly interest me. *But* because they each send me up to five emails a week, I have opted out of their email notifications. If presented with the option of one email a week, I would probably opt back in. Another mistake is to combine different subjects into a single option. For example, as an elite United Airlines customer, I am interested in news on their Mileage Plus program, *but* they have combined such news items with marketing program deals that I generally do not want to see. Again, I have opted out to avoid the avalanche of what I consider spam.

Best practices for creating awareness are to:

- Provide customers with multiple options for how often they receive marketing messages and what kinds of messages they receive.

- Make sure the headline is not misleading—"40% OFF!" may be misleading if it is only on selected items.

- Provide just-in-time notification that is critical to relevance; for example, Foresquare.com notifies the customer of special deals when they enter a store or walk past it.

- Encourage your customers to share their referrals, whether Facebook likes at Chick-fil-A or shared emails with special offers at Starbucks.

- Proactively provide tailored product offerings based on a customer's previous purchases, as Amazon does.

- Properly targeted emails; for example, if your company knows the customer has a dog, cross-selling pet insurance is appropriate and may be a value-add.

The benefits of thoughtful, honest marketing are higher purchase consideration and an image of honesty and caring, as well as no unpleasant surprises later in the CE.

Providing Information to Support Purchase Decisions

As a 2012 *Harvard Business Review* article points out, customers often research products before purchasing them or even approaching the company.[1] Customers investigate partly because they doubt marketing messages and partly because recommendations from friends and reviews on websites can reduce the number of candidates considered. The most companies can do is make their website and marketing material honest, as well as easy to find and understand, and to provide a great CE to minimize negative reviews and maximize positive ones.

Most customers will not expend much effort to log in to a website or read extensive information just to decide whether they might want to buy. The three major company-caused barriers to customers getting decisive information are:

- Requiring extensive information or log-ins before providing answers to simple questions.
- Using ads and websites long on slogans but short on specifics.
- Providing a mass of information that requires extensive effort to digest.

Best practices in supporting customers' information search include:

- Answering the most common questions on the first page of the literature or website.
- Showing the product advantages on a YouTube video of less than 45 seconds.
- Sponsoring customer communities who provide advice and support to other, less expert customers. For example, Intuit sponsors an online community of several thousand CPAs and other users of its accounting software. Scores of expert users provide advice to newer users. Zappos' community provides information on which brands of shoes are sized small or large to reduce returns due to poor fit.
- Encouraging review sites. In Europe, TripAdvisor provides feedback cards to encourage customers to comment to the vendor and post on the TripAdvisor website. This becomes a win-win-win for the vendor, website, and customers.

One emerging challenge in the area of customer reviews is the manufacturing of positive or negative reviews. *The New York Times* reports that there is a cottage industry of companies who will post reviews for $5 each—positive on your site or negative on your competitor's site.[2] A better way to get good reviews is to ensure that your company earns consistently great ratings by delivering an optimal CE, aggressively soliciting, handling, and turning around complaints, and then encouraging customers with great experiences to post about them.

Delivering the Pitch

The pitch is the marketing offer that highlights both the value proposition, including price, and its limitations. It is often the first page of the brochure or the content of the first 15 seconds of the ad. It should be a simple, honest offering. Common mistakes include omitting key facts or exceptions, listing the many limitations using confusing terminology or placing them in fine print or footnotes, and marketing to customers who have existing problems. Misplaced marketing offers can infuriate customers. For example, advertising vacation packages to customers on hold to inquire about their lost baggage is not a smart idea.

The single best practice is to simplify the offer—limit text describing value and price to three or four clear lines. Dan Hesse, the CEO of Sprint, noted at a 2011 conference in Kansas City that when he eliminated 80 percent of the company's marketing plans and simplified those that remained, customers had higher satisfaction, and he was able to close 23 call centers due to fewer calls and problems created by complex, confusing calling plans. A major auto company had a 100,000-mile bumper-to-bumper warranty that led to improved customer confidence and satisfaction. However, the dozen-plus footnotes and exclusions in the warranty still caused a large portion of unpleasant surprises. For instance, the battery was warranted for only three years. Reducing the number of exclusions and streamlining the warranty would have led to greater simplicity and reduced unpleasant surprises.

Other best practices include:

- Highlighting limitations of the product or special offer up front—While the marketing department cringes at mentioning limitations, doing so provides appreciated honesty.

- Asking customers to self-segment on what is most important to them—For instance, Avis asks customers to select their segment, such as cars with the latest technology, a basic car for business travel or on vacation with children, and so on.

Purchasing the Product

Companies should make the purchase experience fast, flexible, and accurate. The most prevalent unpleasant surprises include finding out that items are out of stock after starting the purchase process, confusing forms that result in errors, transaction failures or forms returned due to errors, pressure to buy additional other products, and unpleasant surprises such as extra fees or long-term commitments.

Best practices for the purchase make purchasing fast, easy, and fun. Amazon has developed one-click purchasing. Zappos representatives love shoes and are enthusiastic when describing them and assisting with purchases. Neiman Marcus Direct provides gentle advice on selections of which clothes go together and which do not. Iams service representatives are almost all pet owners who can identify with customers and talk about their dogs. Technology can let them know in advance whether the repeat customer has a dog or cat.

Another best practice is helping customers avoid mistakes. One insurance company identified the most prevalent mistakes agents and customers made when filling out the application form. Going forward, the company put a Post-it Note on the form listing the three most prevalent mistakes. Those three mistakes were immediately reduced by over 50 percent. Likewise, technology that includes edit checks for entered data reduces failed transactions. It is also frustrating when the edit check finds an error, then erases all data and requires the customer to re-enter all the information—and it leads to many abandoned purchases.

Delivering the Product/Service

A successful delivery is predictable and effortless for the customer—and many customers now want instant gratification, too. The key dimensions of delivery are its timeframe, predictability or reliability, and channel.

Common delivery problems include missed service appointments (late or no-shows), failure to provide customers with progress reports and confirmations, failure to deliver the expected product as specified, and last-minute additions to the transaction requirements. For example, telling someone who is refinancing a home mortgage to supply another document two days before the closing is an upsetting fire drill. It happens because the bank staff has a process of just-in-time review of the paperwork and identified a missing item. This process, while convenient for the bank, guarantees angst for a significant percentage of their customers. It also raises bank costs due to the need to reschedule some loan closings.

Best practices include same-day delivery of products, which satisfies the customer's desire for immediate gratification. Companies such as Amazon are now even looking at faster delivery modes such as bike messengers delivering within the hour and the use of drones delivering even faster. Other retailers allow ordering online and pickup at the nearest store.

Customers also desire certainty in terms of when things will happen. Companies are responding by providing more exact times for repair visits—intervals of 8 A.M. to 5 P.M. or even 8 A.M. to 12 P.M. are no longer acceptable. Many service companies are offering guaranteed 2-hour windows. British Gas even guarantees arrival within the two-hour window, or they pay the customer compensation. Companies such as FedEx go further to eliminate uncertainty by means of real-time, online tracking. With its use of mobile and GPS technology, FedEx now allows customers to see where the package is, along with its temperature (if desired). This eliminates most calls about where the delivery truck is and when it will arrive. This value-added function is the center of a FedEx marketing campaign entitled "Be the First to Know!"

Educating the Customer on the Initial Use of the Product/Service

Good education provides just enough relevant information so that customers can begin using the product and get the most value from it. Customers can absorb between three functions without an aid and possibly up to six functions with a quick-start aid. A thick manual beginning with ten pages of cautions is a poor educational approach. While it satisfies the legal and compliance departments, customers almost never get past the first ten pages.

Another poor practice is the auto salesperson who delivers the new car with a rapid-fire presentation on the 30 functions of the ten dashboard instruments, as well as all the other electronic functionality. The same applies to computers and consumer electronics.

Best practices include making education available in small bites, via many channels and formats to accommodate the diverse range of customer learning styles. Welcome packages should be supplemented by offers of verbal or video supplements. Audi recently ran a contest among sales staff to identify best practices and selected the salesperson who asks customers, "What three things do you want to know about the controls right now so you can drive off the lot? I'll limit my briefing to those three items if you agree to come back in two weeks for a more detailed educational session on the items you still need help with or if you agree to view the CD explaining the dash functionality."

Other best practices include:

- Products should be delivered with both printed and electronic quick-start single pages, with references describing how to reach details on other functionality.

- A new user portal on the main company website is especially useful for those who are visiting the site for the first time or who have just purchased the product.

- Because many customers are visually oriented, a series of short videos (45 seconds or less) is effective product education. Honda Canada Finance introduced its website content and functionality to customers using six such videos. Provide simple graphic installation instruction sheets as well as toll-free numbers and links to online videos. Moen Inc., a leading manufacturer of faucets, has found that this approach makes most consumers confident that they can install the product or replacement parts without a plumber, thereby providing greater value and faster installation.

For example, Tesla provides a welcome call before the car is delivered that walks the customer through the website and encourages the customer to view some or all of the 16 videos (totaling 28 minutes) on the functionality of the car. And the software organization Network for Good (NFG) provides both free and for-purchase software to nonprofit organizations to process donations. NFG monitors its customers to see when they start using their new software. If the customer has not used it after two weeks, the company emails and calls clients to identify why they are not using the software and offer technical support to assist in implementation.

Using and Maintaining the Product/Service

As customers become more familiar with products, they need help getting the greatest possible value from it, as well as assistance with maintenance and billing. The usual unpleasant surprises include early failure of a product, inability to use advanced product functionality, or higher than expected bills for service. The failure can be avoided with technology. For example, devices like Xerox machines automatically phone home when sick before they actually fail.

The same approach to monitoring product startup, followed by the just-in-time education mentioned in the previous section, can be used to ensure a customer's progress. For example, most customers know how to use only a third of the functionality resident in their auto navigation and entertainment systems. The manufacturer can monitor what is not being used and, using customer demographics and psychographics, suggest what additional functionality would give the most benefit. The email or phone message suggestion could come with three 45-second tutorials on how to use the functions, as well as links to more detailed explanations. When customers use more product functions, they perceive higher product value.

Building a Flexible, Customer-Focused Culture That Delivers DIRFT

People, processes, and technology must be properly tuned in nine areas to create a flexible, customer-focused culture.

Clear Brand Promise

A clear brand promise honestly communicates the key aspects of the value proposition. For instance, HSBC is the "World's Local Bank" and "You're in good hands with Allstate." At FedEx, "The World on Time" sets a simple expectation of reliability for both customers and employees. The brand promise is the foundation, which directs the people, processes, and technology in the next eight areas.

Clear Accountability for Delivering the Brand

A key part of being customer focused is universal accountability for the fulfillment of the brand promise. Organizations demonstrate accountability for brand-aligned service in four ways:

1. Formalized structure using a department or cross-departmental team to ensure that the organization delivers the brand's message in all communications and champions the brand in all operational service design meetings.

2. Leadership that leads by example as well as in communications, evaluations, and individual actions. For example, at the Walt Disney Company, all managers spend at least one day a month working with the public; they must also walk the talk and practice what they preach by, for example, picking up and disposing of trash whenever they see it.

3. Peer accountability requires that employees hold each other accountable for "living the brand." Disney and Starbucks staff are encouraged to gently counsel coworkers who forget to wear their name badges or act according to the companies' values.

4. Employee empowerment, meaning that employees are authorized and encouraged to actively and creatively help deliver the brand message. The Disney staff is provided Make the Experience Magical cards that they can give to guests. The cards can be redeemed for specific items (e.g., a box of popcorn or a balloon) or privileges (e.g., skip to the front of the line) at the employees' discretion. The cards are tracked so that managers can praise the employees for taking the initiative to give a card to a guest.

Focused Values Translated into Operation Guidelines

Customer-focused organizations have a few catch phrases that epitomize the company's credo and provide a framework for putting the brand into practice. For instance, at BB&T Bank, values include reality, reason, honesty, and integrity. Employees latch onto the phrases and create service experiences around their meaning. Decision making in the organization is driven by these values. Interestingly, BB&T's value statement says it is OK to be emotional, as long as emotions are expressed rationally and logically, and they reinforce the best decisions. At Johnson & Johnson, part of the firm's credo states, "[O]ur first responsibility is to the doctors, nurses and patients ... and all others who use our products and services." This simplifies and speeds decision making during a crisis.

Formal Processes for All Key Activities

The design of every process that directly or indirectly impacts a customer must focus primarily on creating the ideal CE. Process mapping should be used to highlight disconnects and unnecessary demands made on the customer to see if they are really necessary. For instance, in one insurance company, the process map showed five places where the company recontacted the customer for additional information or clarifications. When the diagram was shown to executives, they were able to quickly consolidate and/or eliminate four of the five steps.

Measurement and Feedback

Customer-focused organizations use both direct and indirect descriptors of the CE to provide insight into performance and ensure that the company's CE tracks with its brand promise. Direct data sources include customer satisfaction measurement, complaints, market research, social media, and VOC processes. Indirect sources include quality measurement, operational data, online review sites, and continual feedback on employees. In many cases, the indirect data sources are timelier than the customer data sources because the company's operational system is aware of the problem before the customer becomes aware of it. Chapter 6 is devoted to building and using measurement and feedback processes.

Ongoing Communication to Customers, Channels, and Employees

At organizations with a great CE, the CEO communicates frequently with everyone to demonstrate and reinforce brand principles. In addition, training, newsletters, and guidance go out to everyone in the organization, including those at the lowest levels, and to customers and distribution channels as well. One company posted the newsletters several places in the restrooms (which all employees use) and entitled it, "Flush Facts." These communications are typically steeped in storytelling. Disney recently published a video on how a little girl's lost princess doll was not just returned to her but accompanied by princess characters carrying a scrap book (assembled by other Disney members) with pictures showing the exciting day the doll had meeting other princesses. This video not only commended the employees but showed everyone how a mundane action of returning a lost item could be made into a remarkable CE.

Intentional Emotional Connection

Organizations should plan to create emotional connections with customers. The two parts of this connection action are creating opportunities to connect and empowering and training employees to make the connection happen. In some cases, connection can even be achieved with technology. This connection is a function of an empowered, customer-focused culture.

The biggest challenge of DIRFT in creating connection is deciding whether to create transactions that naturally allow connection or to try to create connection during mundane routine transactions. An example of the first approach is to express concern and ask whether everyone is OK when taking a claim call at an auto insurance company. The second would be a server at a restaurant asking how the guests' day has gone and if the meal is a special occasion before moving on to take drink orders and describing the specials. We will address this important area in more detail in Chapter 9.

Employees Who Deliver the Brand

Organizations must hire people who fit the brand image and possess the right psychological attributes. Companies can teach people the necessary skills but cannot change their basic personalities. Levi Strauss's customer relations department has a hiring profile that targets confident, outgoing, and "real" individuals who fit the company's core brand values. Southwest Airlines specifically looks for employees who are animated, warm, and fun. Whole Foods Market recruits new employees from customers in the coffee bar and other departments in order to ensure its staff has a passion for natural foods. The theory is that people who enjoy the atmosphere at the front of the store are more likely to enjoy working there.

Customized Brand for Market Segments and Geographies

Customer-focused organizations plan globally but maintain the flexibility to meet regional and market segment needs. For example, Bank of America varies its service specifications based on customer expectations in different areas (e.g., urban versus suburban markets), as well as geographic differences (e.g., New York City versus Southern markets). The retail store Target varies its product mix by region and ethnic makeup of the customer base. Frontier Communications employees answer the phone with their full name and their office location to stress their local connection with the community. A major Russian bank serves coffee, tea, and snacks in rural village branches because customers value social interaction over speed and use the wait in line to catch up with their neighbors. However, in Russian urban areas, efficiency and speed receive the most attention.

DIRFT with Flexibility

Delivering on the promise every time with flexibility requires tailoring your value proposition (statement of why customers should buy your product) and the actual product, as well as the delivery and user experience, to the individual customer's needs. The challenge is building in flexibility while meeting legal, compliance, and process requirements. For example, the head of CE for a major financial services company mused to me, "I wish we could get the front line to break the rules for good customers." I suggested that compliance and legal constraints would preclude breaking the rules *but* that, if the rules were drafted with flexibility at the issue level—what I call flexible solution spaces—then the customer service representatives (CSRs) could break the rules without breaking the rules.

The solution is to anticipate most of the variations that the customers might request, to design the CE process, and to equip employees to vary the product and delivery, within limits, to meet those disparate needs. For the product and offer, you can plan most flexibility in advance. In the delivery and use phase, you have to ensure that the company's processes, technology, and people are all flexible enough to deal with the prevalent unexpected customer request or external event. Once customers have bought the product and begun to use it, their needs or perceptions may change. Note that flexibility usually varies with the value of the customer. For a gold customer with a cancelled flight, the airline employee might put the passenger on a competing airline to get them there on time while a nonelite customer would be offered only the next available flight.

Creating Flexibility in Your Product and the Marketing Offer

Your product and its marketing offer must allow for common customer preferences, including product attributes (e.g., color, amount of digital storage), payment mechanisms, and length of commitment, as well as customer familiarity with your brand and overall product category. For example, savvy staff of computer and wireless stores will first ascertain the general level of the customer's technological sophistication and how they want to use the product prior to describing the product attributes. Staff members tailor the recommended product and their pitch and explanation toward attributes the customer will understand and care about in such a way as to reduce uncertainty.

My wife and I were recently in southern Italy and were struck by the fact that many of the restaurants have no printed menu. The waiter simply suggests what he would like to serve you, often without mentioning prices. This is the custom. But because we were not used to this kind of offer, we always wondered how much the bill would be and worried that it would be outrageous. At one small restaurant in Ravello, the waitress briefly described the seven offered entrees and included a price range—sort of a verbal menu. After that, we were perfectly happy with no printed menu because this waitress educated us and reduced our uncertainty.

Creating Flexibility in Product Purchase and Delivery

Flexibility in purchase and delivery primarily addresses the range of communication and delivery channels and speed of the delivery.

- When customers are online, recognize each from a single identifier and prepopulate much of the screen for them. Allow the customer to make changes without going back to the beginning; if customers change something, don't force them to reenter everything else.

- Make it easy for the customer to say no to extra services; people hate having to search for the little "no" box hidden at the bottom.

- Allow the customer to test-drive the product without obligation. Communications companies allow customers to try a channel free for a month. Car dealers allow the customer to take the car home for a weekend.

- Some catalog companies let customers try clothes on virtually, creating an electronic mannequin that looks like the customer.

Creating Flexibility During Ongoing Use

Flexibility in use begins with the selection of startup educational channels. The Audi salesperson I described earlier asking the customer, "How much education do you want now versus the CD, the online tutorials, or a return visit in two weeks?" is a good example. The customer was given four options to choose from.

Next, give customers a range of alternatives for using and maintaining the product or service, including ongoing education, support, and billing options. One best practice is proactive education on how to get more out of your product. For instance, Apple charges $99 for a year of educational sessions on using a new product. They do not remind customers to come in for more education, but if they do, customers would buy more upgrades, products, and services.

Establishing the Right Metrics for DIRFT

Perfect DIRFT requires metrics for each phase of the end-to-end CE. They should especially address the marketing and first product use activities that are seldom measured except in the auto industry, which has worked hard to eliminate the negative impact of the car purchase experience. The metrics should both measure the degree to which customer expectations were fulfilled without problems, as well as whether the company delivered emotional connection and extra value.

Four sources of metrics should be used to monitor success in DIRFT processes: (1) customer comments (complaints on unpleasant surprises and compliments of good experiences), (2) surveys, (3) operational data, and (4) employee input. How each type of metric can be applied to some of the more challenging activities of DIRFT is outlined here and discussed in detail in Chapter 7.

Metrics That Identify Problems in Each Phase of DIRFT

Many of the needed quality metrics are routine, but some phases require creativity. Four sets of metrics highlight problems in each phase:

- *Complaints*—These are the most visible flags. However, customers seldom complain about marketing and sales issues unless feedback is explicitly requested.

- *Survey data*—This must be systematically collected, ideally using a single satisfaction question followed by an open-ended description of the reason for a rating level. There should also be a more comprehensive annual study of the end-to-end customer experience that enables the understanding of the key drivers of customer loyalty, their impact on WOM, and the relative priority of problems encountered.

- *Operational data*—This is the most reliable information on process failures. The challenges are that not all disappointments can be noted by systems and that getting at the data in your system can be difficult. My advice is to start with metrics on the most prevalent and damaging problems, such as missed deliveries, returned products, and service charges that will be a surprise to the customer, and then slowly expand the reporting as systems are revised and updated.

- *Employee input*—Your customer contact employees are one of your best sources of information on what is working and what is not working. Hold a weekly meeting for frontline employees to give input on their most frustrating customer interactions and the underlying causes.

Metrics That Identify Connection and Value-Add

Because value-add and connection are in the eye of the beholder in many cases, customer feedback is the most effective source for these metrics. The three best sources for measuring connection and value-add are surveys, employee input, and compliments.

- *Surveys*—Although customers are less likely to complete a survey when they are happy than when they have encountered a problem, a very simple, one-question survey can be effective, especially online. Once people have given you a numerical score, they are quite likely to provide a short text statement about their reasons. If the transaction was important, either financially or emotionally, you may well get a full paragraph of description.

- *Employee input*—Employees will be happy to relate actions they have taken that resulted in compliments or happy customers.

- *Compliments*—Solicited and unsolicited compliments are very helpful. The "recognize our staff" coupons given to customers seldom work, though. However, a "feedback on our employees" link on a frequently visited part of the website, such as reservations or account maintenance, generates lots of feedback.

The Role of Technology in Collecting the Data to Support the Metrics

Your systems and technology are continuously collecting data on interactions with customers. The challenge is that data in the system is not actionable. Websites, phone trees, CRM systems, and speech analytic packages contain a wealth of data, but you, as the CE manager, must tell the systems managers what you want to know. Your guide should be the charts of the key customer-facing processes discussed at the beginning of this chapter that will highlight all the possible problems. From these charts and your survey and complaint data, derive the top five events you want to prevent, and task your technology peers to find you the data that best describes those events' frequency and cause. Get some quick victories from those and then move on. My key point is to pick your battles and get a few wins before demanding more data.

GETTING STARTED

Questions to Ask Yourself About Your DIRFT Processes

1. Do you have descriptions of your marketing pitch, purchase, and product delivery processes, and have you highlighted the unpleasant surprises during each?

2. Do you have a customer-focused culture that fulfills the brand promise?

3. Have you built flexibility and intentional emotional connection into your processes?

4. Do you know how many unpleasant customer surprises the existing processes are causing?

5. Are you measuring DIRFT success end to end for each activity?

....

KEY TAKEAWAYS

- DIRFT must cover every activity from awareness and marketing to startup and ongoing maintenance.

- People, processes, and technology must be combined in nine areas—from brand promise through communication and hiring—to create a customer-focused culture.

- Flexibility must be built into almost every part of the DIRFT processes using technology guided by the ideal CE process map. Staff must be provided with flexible solution spaces for key issues.

- Emotional connection is not needed everywhere but should be intentionally built into one or more phases of the CE.

- Metrics must be developed across every phase of DIRFT, including marketing and product use. Technology is the best source of many of these metrics.

Notes

....

1. Patrick Spenner and Karen Freeman, "To Keep Your Customers, Keep It Simple," *Harvard Business Review*, May 2012, p. 109.

2. David Streitfeld, "Race to Out-Rave, 5 Star Web Reviews Go for $5," *New York Times*, August 20, 2011, p. A1.

CHAPTER 5

Keep Every Door Open

Assuring Multichannel Access

A fellow consultant told me of a seminar he led for executives responsible for the customer service operations of a large business-to-business company that provides human resources services. The company has an award-winning operation but is seeing warning signs of changes in customer expectations. Part of the discussion was about what level of customer service should be provided through social channels.

As they talked, the consultant went to the company's Facebook landing page (projected to the front of the room) and heard a collective gasp from the group. Next to a marketing message in the left column was a scathing post from a customer befuddled by a simple password reset issue. The message, detailing the customer's grievances in trying to contact the company and get the problem resolved, had been on their Facebook landing page for over 48 hours. This led to a rapid-fire series of questions among the group. What happened? Why could the customer not get through? Did he have the right contact info? Why was the marketing department not monitoring these posts? What was the process for responding to these types of customer messages?

We are entering a new era of customer access, characterized by ever more cynical consumers who are too busy to use a ponderous complaint process but who are willing to go public if not quickly satisfied. Further, the number of channels by which customers can communicate with companies is multiplying. To maximize the percentage of customers who give you a chance to solve their problem you must convince customers to use your service system before going public, as well as open up new, effortless channels and deliver consistent, effective responses across the full range of contact channels.

In this chapter, you will learn:

- How to encourage valuable contacts from customers with questions and problems by breaking down the perceptual barriers to contact.

- The factors that affect the proportion of customers needing help who will actually contact you, thereby creating service workload.

- The components of an effective customer access strategy and how technology can help facilitate access.

- The metrics that provide an actionable picture of accessibility.

The third section of this chapter, on the components of an effective access strategy, is heavily based on Chapter 3 of Brad Cleveland's book, *Call Center Management on Fast Forward*.[1] I highly recommend it. Although *Fast Forward* is focused primarily on telephone and electronic interaction, much service continues to be given face to face and by field sales reps in B2B environments; I will add suggestions in those areas.

Encouraging Valuable Contacts

Valuable contacts are those that, if handled well, will increase loyalty, sales, and WOM. Value is produced either directly or indirectly by creating a positive experience that later turns into fewer problems or more sales. Chapter 4 showed that it is financially more profitable to receive and handle contacts from customers needing assistance than not to hear from unhappy customers. This approach differs from some books suggesting that the best service is to not have the need to give any service. Although giving no service does translate into re-duced cost in the present, it can lead to future problems and disloyalty due to no education as well as missed chances to create a positive, memorable, and hopefully inexpensive connection.

The key to encouraging valuable interaction is breaking down the four perceived barriers identified in Chapter 1: (1) It is hopeless and will do no good; (2) it is too much hassle; (3) conflict and retribution are feared; and (4) a preferred communications channel is lacking. Strategies for addressing the barriers vary by when the customer encounters a question or problem. If customers are at home on their computers, they should be presented with a message encouraging contact via the website email tool. If they are in a car dealer or restaurant, they should be encouraged by both signage and staff demeanor to talk to a staff member or the manager. However, some customers may be more comfortable complaining via another nonconfrontational channel such as webchat or an 800 number rather than face-to-face or email. Others may prefer recording a message to management via their mobile phones. Some customers may want to switch channels midtransaction.

For example, if you check on a package delivery online and find that it has not been delivered on time, you may then want to talk to a human. If customers encounter a barrier to easy contact at any point, they may well abandon service and remain dissatisfied. You should review all marketing messages and service touch points to see if the company has erected any of the following barriers to contact. The barriers are addressed in the order of prevalence, as measured by the *2013 National Rage Study*.[2]

It Is Hopeless to Complain—Complaining Will Do No Good

The *2013 National Rage Study* confirmed what almost all of my other research has found: The single most prevalent reason for not complaining is the belief that it will do no good. The customer must be convinced that the company wants to hear problems and will fix them. The message to customers should be that the company wants to hear about problems. Ideally, the message should provide examples of those types of complaints that customers might encounter (or be reticent to voice) as further encouragement to communicate. For example, a hotel chain had a guarantee for room service—"It's On Time or It's On Us!" The company knew that a prevalent problem with room service meals was longer than expected delivery time but that customers often would not mention problems because nothing could be done once it was late. The solution was a message highlighting timeliness countering any perception that the company did not want to hear about the issue.

The basic rule to defeat this barrier is to have the message in front of the customer exactly when they encounter the problem. BOSE Corporation places a message card in the box with headphones that says, "We're here to HELP! Contact for assistance with setup or troubleshooting." Further, if you send a complaint solicitation message in customers' statements of their account status or bills, the message should be prominent on the first page, not buried in the print at the bottom or on the back of a paper bill. On webpages, the "Contact Us for Help" message must be at the top of the page in large print, not hidden below the fold at the bottom.

It Is Too Much Hassle to Complain— Complaining Will Take Too Much Effort

In most environments, the perception that getting assistance will take too much effort is the second most prevalent barrier to customers contacting a company. This is especially true if the customer has a question on only one aspect of the use of a product or about a minor charge on their bill. Customers, who have very busy lives, often assume that they will have to wait in queue on the phone or in line at the service desk. Further, they fear that they will be required to provide extensive proof of purchase or other paperwork. Therefore, they say to themselves either, "It is not worth my time" or "I will do it later when I have time." In the majority of cases they never take action.

Best practices are to make complaining easy with no waiting and to require a minimum of information and effort to obtain service. Both of these facts should be conveyed in the complaint solicitation message. An example of an effortless channel is a mobile app, 45sec.com, which allows the customer to simply touch and record on their mobile phone—speak about their complaint for up to 45 seconds—and hit submit.[3] The app software identifies the appropriate recipient of the complaint (both company HQ and local manager) and transmits the complaint to the company within 90 seconds. In another example, LL Bean is known for its 100 percent satisfaction guarantee. The company stresses on the Web home page that there is no time limit for returns and that no authorization is needed from Customer Service. It says, "No Fine Print—No Restrictions." Although some finance executives believe that soliciting complaints will increase fraud, I have found general agreement among service executives that 98 percent of customers are honest.

Fear of Conflict or Retribution

Customers fear that a complaint will result in employees arguing with them or that the employees will feel personally criticized, possibly leading them to strike back at the customer. For example, when the employee asks a question in a directive manner such as, "Is everything delicious?" customers are afraid to contradict the employee by saying the food is just so-so. Worse, they may fear future retribution from the employee, especially if the employee can affect their well-being. Think of the server who controls the quality of the food or the appointments secretary who controls access to a doctor's appointments. In a face-to-face environment, an employee breaks down this barrier by making eye contact when asking whether everything is OK or if the food is to your liking.

The solicitation message and the demeanor of the frontline staff must both convey that any dissatisfier is of interest to the company. This is especially true in B2B environments where customers are afraid to complain about sales reps and service technicians because they know they will need to do business with them again sometime soon. Therefore, alternative channels of communication to headquarters or management must be available. A technology company has required that each employee's email signature block have the phrase, "How am I doing? Tell my boss!" with their boss's email address. This not only encourages complaints, it also provides a channel for compliments, which are seldom submitted because customers do not know whom to contact.

Lack of an Easily Used, Preferred Channel of Communication

Even if customers are motivated to complain or seek assistance, they are often stymied by the unavailability of channels that they are comfortable using. An acquaintance found some small, hard, black lumps in her child's cereal. She first emailed the company and, after waiting for 24 hours for a reply, then called their 800 number at 7 P.M., only to find that they were closed for the evening. She then posted a digital photo on their Facebook page, labeling it suspected rat droppings. She went social with her complaint only after being stymied by the nonresponse of the company's email process and then the telephone service center. The channel that the customer prefers should be offered and responsive at reasonable times when the product is most likely being used. At a minimum, this customer should have received an email reply within 24 hours. Most customers now expect replies in significantly shorter time frames.

The customer should be provided with all the possible channels so that they can use the channel preferred at that time. The website, as well as printed company literature, should provide chat and an 800 number besides the email address. In addition, easy transition between channels should be facilitated. For instance, if I am tracking a package delivery on the website, I am happy to use the Web self-service until I find that the package has not been delivered on time. Now I want to talk to someone. A click to call or a call-me button should be prominent on the same page where I learned of the delivery failure. If the technology is really effective, the call-me message software that connects the customer will also provide a tracking number and customer's identity to the service representative who handles the call.

Factors That Drive Contact Workload

The volume and mix of customer contacts are a function of four factors, as illustrated in Figure 5-1. The fundamental factor that drives service workload is the customer's need for information and service. This workload is caused by customer problems or confusion resulting in customers deciding whether they need assistance. If the product causes no unpleasant surprises or questions, there will be a low level of service workload.

Figure 5-1

Factors Driving Contact Workload

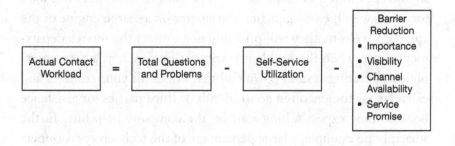

The degree to which these problems and questions actually become service workload is a function of the other four barrier factors: importance, visibility, channel availability, and the service promise. Each of these four factors is more or less a barrier that will reduce the volume of communications actually submitted by customers to the company. For example, if the preferred channel of communication is the telephone and the call center is not open on weekends, there is a high probability that problems encountered on the weekend will not be submitted for assistance.

Volume of Customer Need for Information and Service

The primary factor that drives the customer to contact and request assistance is the customer's need for information or service both before and after purchase. The incidence of problems and needs will vary according to the complexity of the product. Often, customers may do a general Web search and/or check the company's website prior to visiting, calling, or emailing. The more complicated the product and its directions for use, the higher the number of questions and problems.

Importance of the Question or Problem

When a problem is encountered, the issue is minor, and involvement is low, there is a good possibility that less than 5 percent of customers with the question or problem will search the website and/or contact the company. If the problem is serious and/or will cost the customer significant money, as many as 50–70 percent of customers will look for an answer. If customers find the answer on a search engine or the company website, they will probably not contact the service department. Even when the problem is very serious, such as a new smartphone operating system download that erases all your contacts and calendar, customers often go to friends or third parties for assistance because they expect a long wait on the company help line. In the smartphone example, a large percentage of the tech-savvy customers will go to the online community or the company website prior to actually calling or trying to chat with a support person.

Visibility of Preferred Channel

The physical visibility of both the customer contact solicitation message and the various communication channels are critical to the volume and mix of contacts that a service system will receive. I recently examined a soft drink can in South America to find a customer service contact. I had to carefully study the finely printed paragraph on ingredients, place of manufacture, and so on—finally in the eighth line of text I found the statement, "Satisfaction guaranteed, call xxx-xxxxxxxx." Having the customer contact information but hiding it in the midst of text, or putting it on the bottom of the package or the back of the claim form, is almost as unhelpful as not providing the contact information at all.

Two examples illustrate creative ways to make access to service effortless. A hospital in Maryland advertises on the radio that its emergency room (ER) waiting period is usually short and encourages consumers to check the real waiting time via a mobile app. Here the company tries to proactively anticipate the customers' need for more information in an easily accessible manner. The communication is a good investment, given the revenue implications of the average ER visit. In the second example, Rosetta Lue, Philadelphia's Chief Customer Service Officer, has trained over 1,000 neighborhood liaisons to solicit problems from neighbors and enter service requests directly into the Philly311 computer.

I have executed a number of experiments with companies in which the size and placement of the text messages were varied on packages and websites. The physical location of the message can influence the number of contacts received by a factor of at least ten and sometimes 50. Therefore, if you are not getting many complaints or support calls for a complex product, you should examine the visibility of the communication channel to determine whether customers are effectively being made aware of the contact information.

Channel Availability

Channels may be available 24/7 or only during certain hours for certain types of services. If the channel is not available when the problem occurs, the customer will not bother to wait until the channel becomes available. A good example is a soft drink company that did not have their hotline open on weekends or evenings. A problem with contamination became disastrous before being addressed, in part because many customers encountering a strange taste on the weekend or evenings (when a majority of the product is consumed) simply discarded the product when they found that the hotline was closed after 5 P.M. on Friday rather than keeping it until they could call when the hotline was open. The soft drink company's director of quality estimated that they would have received at least four times as many calls and would have recognized the problem much faster had the hotline been available when the bulk of the product was consumed.

Another barrier is wait time in queue, whether physical or on the phone. If customers encounter long wait times in line or on phone queues, they will abandon their efforts unless offered a specific service time provided by virtual queue technology. This function notes the customer's phone number and calls them back when their number comes to the head of the queue. Most companies report that the virtual queue dramatically reduces abandoned calls and significantly improves customer satisfaction.

Availability in a face-to-face setting can also be affected by physical factors. For instance, if there is a line waiting for the service person in a department store, the customer will probably decide not to wait to ask whether the store has the pillowcases in blue. If the counterperson at a fast-food restaurant does not make eye contact, the customer is less likely to ask questions or voice dissatisfaction. Even security windows decrease communication by making it hard to see and hear the person on the other side at banks and ticket booths.

Service-focused retailers and hotels now have the staff come from behind the desk to interact while making eye contact or are experimenting with facilities without counters. Employee behaviors signal availability or lack of interest in helping. Three inappropriate behaviors are:

1. Folding your arms in front of you, which signals resistance.

2. Staying seated when a customer approaches you or not sitting if the customer sits so that you are both on the same level.

3. Using a tired or uninterested tone of voice, creating what is called verbal/nonverbal mismatch that indicates you really are not interested in helping.

In retail environments, staff must be visible and available. Here, *available* means without a line, not talking to a fellow staff member, not stacking products on a shelf or doing paperwork. Staffers who look busy and do not make eye contact or acknowledge the customer are, in effect, communicating to the customer that they are not available. Further, if they are busy doing something when they can see the customer, they are, in effect, telling the customer that their current activity is more important than the customer. When employees see the customers, they should acknowledge them by at least nodding and smiling even if they are in the middle of helping another customer. In retail service environments, these behavioral cues can trump all the signs in the world saying, "We want to help."

Service Promise

The service promise addresses exactly what is promised by the solicitation text. A statement of "Questions or comments?" is less motivating than "Satisfaction guaranteed," which, in turn, is less motivating than "For assistance with setup or troubleshooting." The more specific the message is, the more contact you will have, especially on the issues named in the statement.

Combining the Factors to Estimate the Volume of Contacts

The procedure for estimating overall volume is first to estimate the total number of customers who will need assistance and then to estimate how many will actually run the gauntlet of the other factors of importance: visibility, availability, and service promise. The best approach is to look at a comparable existing product and see how many customer contacts that product is producing via each communication channel. Next, note how the new product differs in complexity, need, labeling, and potential problems in order to estimate how many customer contacts the new product will generate. Estimating how many contacts you will get in advance of actually launching a product is more art than science, but extrapolating from previous products should allow you to be accurate within 20–30 percent.

You can also learn a great deal from past mistakes. A cat food manufacturer accidently mislabeled a product but did not identify the problem until after it was shipped. They knew how many units were shipped and also how many calls they got for that problem. Thus, the company had an empirical "stake in the ground" as to how many contacts a particular type of problem would generate.

Planning Customer Access— More Complicated Than It Seems

Most companies assume access is answering the phone quickly or having staff available. The company needs to train the people with the process and technology of the service system to assure that access does not create more customer frustration. Access must be placed within the context of the overall customer experience plan, as described in Chapter 4. There are ten key components of the Customer Access Strategy as outlined in Cleveland's book *Fast Forward*, and summarized in Figure 5-2.[4]

Each of these components will be described, along with the related most prevalent mistake people make. (For the details in implementing each of these areas for contact centers, see Cleveland.) For implementation in face-to-face environments, I've added prerequisites and mistakes in each area.

Figure 5-2

Components of Customer Access Strategy

1. Segments	6. Routing Method
2. Workload	7. Resources Required
3. Communication Channels	8. Information Required
4. Hours of Operation	9. Analysis and Improvement
5. Service Level	10. Strategy for Introducing New Services

Segments

Customer market segmentation clusters customers with similar needs or characteristics to allow more targeted marketing messages or focused service. Segments should identify the types of customers who receive different levels of service so that the access process can flag and route the customer to the staff that can best help them efficiently. Segments inform the service system about the customer's general profile at the start of the transaction. This knowledge becomes especially important when different skills or levels of expertise are needed to service various segments.

I generally recommend limiting segments to three or four designations that are easily discerned by frontline staff. For instance, in an airline, there may be general, leisure passengers, elite frequent flyers, customers who paid for business class, and super elite passengers. The reason for segmentation is to give staff guidance on where they have more flexibility in handling customer issues, such as in rebooking flights or seat selection. The prerequisite is that the staff be able to easily identify the customer's segment, by either the phone number or frequent flyer number, from the ticket, or by some other easily discernable indicator. Segment should not determine the basic approach to how customers are treated; they all should be treated with courtesy and respect. The flexibility of the response should be driven by the segmentation. This decision making may also be affected by the understanding that the platinum customer is probably giving you all the revenue that they generate while your silver customer could be a competitor's gold customer—meaning you could gain the most incremental revenue from the silver customer with great service and demonstration of expertise.

In a retail setting, the most prevalent mistake is guessing customers' segments based on their appearance. It is better to treat everyone as a gold customer, at least until you hear their answer to the question, "May I help you?" Customers who buy from a competitor are still worth the effort. In fact, if you impress those customers enough, they may give you a chance to win more of their business.

Other useful ways of segmenting customers could be by product preference, such as preferring large, high-tech rental cars versus simple compact cars, or by technological savvy, such as novices versus techno geeks, which could determine the CSR's approach to providing technical support.

The biggest mistake most companies make is creating too many segments or introducing multiple segmentation schemes into the service system, forcing staff to slow down while they determine which segment the customer is in. It also increases the probability of misclassifying customers and therefore annoying them. Keeping it simple is the best strategy.

Workload

The workload component describes the volume and types of customer contacts that will approach each communications channel. The first task is to forecast the workload by communication channel and by staff hours to ensure you have enough employees to handle the workload. The second task is to forecast the mix of workload (product purchase, simple support, complex support) to be sure you have the resources with the right skill set needed to satisfy the customer. Poor forecasting has been the primary source of many customer service disasters. Cleveland's *Fast Forward* provides a great guide for how to successfully forecast workload.[5]

The biggest mistake companies make is to forecast workload on a monthly or yearly basis. Every month and every year has seasonality (variation by day of week or day of month), as well as special markdowns and sales, which bring in many more customers. In addition, new product introductions and quality problems will also create spikes in customer contacts. The really good companies have a dynamic forecast of workload by communication channel at least daily. The best companies use technology to reforecast hourly and take into account both internal and external events that can impact workload and resources needed.

Communication Channels

Communication channels define the full range of communication mechanisms that customers can use to get service. Usually this will include Web self-service, mobile app, text, chat, toll-free number, and for retail/service offerings, face-to-face and self-service kiosks and automated tellers. I differentiate between mobile app and Web self-service because many companies simply put their webpage in the mobile app and most webpages are impossible to read or use on a smartphone screen. So far, social media is not a major primary service channel but must be considered. In B2B environments, most people contact field sales reps first, but many companies have not thought through what the customers should do outside of regular business hours. Many small and medium-sized business owners do much of their decision making and have questions for a company they are doing business with in the evening and on weekends. If the account rep is not available, who should get the call or email?

The biggest mistake most companies make is not prominently offering links to all other communication channels (to allow easy access to the alternative channels) and not offering the same extent and hours of service across all channels. The second mistake is not thinking through how customers who usually go to the store or who use the phone, email, or social media will get service when the store or contact center is closed.

Hours of Operation

The hours of operation for a phone/Internet-based service system are very different from the hours of operation for a brick-and-mortar service system. The service system has to be available when the customer needs it. How do you determine when the system is needed? When is the customer using the product?

If your service system is not available when the product is being used, you will not hear from customers and will not learn of their problems and/or make the sale. Further, social media must be monitored and triaged 24/7 for ticking time bombs. The biggest mistake most companies make is assuming that the customers' times of use or need for support is congruent with the company's office hours.

Service Level

Service level is the speed of access to the service system and applies both in a phone and a retail environment. It is based on an estimate of how long the customer will wait in a line or in queue on the phone. *Fast Forward* makes the point that in retail settings customers can see the line moving, whereas when on hold in a phone setting, they cannot.[6] Cleveland provides seven tolerance factors that apply well to both phone and retail settings: (1) level of motivation, (2) availability of substitutes, (3) the competition's service level, (4) expectations, (5) time available, (6) cost of service, and (7) external factors such as weather and current news events.

As mentioned earlier, the new technology of virtual queues has revolutionized wait times by eliminating the need for customers to wait on the phone for their calls to be answered. Customers can now be told the wait time is six minutes and asked to press one if they would like to be called back in six minutes. With this approach to phone queues, customer satisfaction is quite high—as long as you actually call the customer when promised.

The biggest mistake most companies make is placing a labor cost on each minute of labor when the CSR is on duty but assigning no cost or revenue impact of abandoned calls or customers who walk out of the store without buying due to a line at the cashier's counter. This results in the cost-benefit analysis always being skewed toward cost saving versus an enhanced CE.

Routing Method

This part of the plan defines how contacts from each customer segment flow through the access system to the service system that is designated to handle specific types of customers with a particular type of issue. Routing can include the use of automated systems, such as automated call directors (ACD) and touch-tone– and voice recognition–driven IVR systems for routing phone calls. In many cases, the customer segment (e.g., gold, silver, or bronze) can be recognized via an automated number identification function in the ACD that recognizes the special segment phone number. Emails and chat requests are routed to the appropriate email or chat queues based on customer and issue classifications coded into the email or chat template. The access plan specifies which customers are sent to which group of service representatives or which part of the self-service process. Many CRM systems have become sophisticated enough to allow the blending of email and phone calls so that the same staff member may get two calls to handle, followed by an email. Also, be sure to consider virtual queuing for when waits exceed three minutes.

A big mistake is creating a routing plan that has a high potential routing error built into it. Sources of misrouted calls include complicated IVR menu options that confuse the customer into selecting the wrong option, speech technology that misunderstands the customers' request, and customers who call from phone numbers other than those they are normally associated with. If a routing mistake is made, customer satisfaction is severely damaged, and service cost rises. My research in several companies found that when customers are misrouted, they are doomed to be transferred at least two more times before getting to the right place and are often forced to wait in a new queue or, worse, get cut off. Therefore, you should minimize routing errors via a very simple routing scheme and, for phone calls, provide the IVR menu wherever you provide the 800 number so that the customer knows which option to select.

Resources Required

Resources include the service system employees and the technology required to handle the full range and volume of issues presented by the workload. Human resource requirements are driven by two factors: the forecasted workload and the cost of the resources. The human resources requirement is further complicated by the diversity and varying complexity of issues presented by customers and the sophistication of the technology supporting their response. In many cases, staff requirements are divided into generalists and specialists, further complicating forecasting and scheduling.

The technology requirement includes the CRM and Knowledge Management Systems (KMS) supporting the self-service and Web functions, as well as the human representatives who respond to the service requests. Further, these must be linked to the databases supporting new product offerings (for example, smart meter billing systems in a utility) and new channels such as mobile apps. The linkage must now be based on the same customer identifier across all customer-facing systems. Technology now should also include speech analytics and workforce planning systems.

The two biggest problems are caused by failures to effectively forecast workload volume and the failure to keep the information content of the KMS technology up-to-date. The forecasting failure causes a mismatch between workload and resources. Also, when under time pressure, the staffers tend to give less complete answers, fail to emotionally connect, and rush customers off the phone. Even though the damage is obvious, many organizations forecast only monthly or weekly rather than multiple times per day. If the KMS information databases supporting response and the self-service Web functions are not up-to-date, the ability to respond starts to crumble and callbacks increase.

Five types of technology have improved performance in this area. The first is workforce forecasting and management software. Such software is only as good as the data input and the diligence with which the manager uses it. If the workload history is entered only weekly and a new forecast produced monthly, you will often miss the mark. Virtual queues can smooth out peaks by delaying the handling of customer calls by 5–15 minutes with very little impact on satisfaction and often with a positive impact on productivity, assuming the customers are called back on time. Third, if an unexpected workload peak is encountered, standby home-based CSRs allow companies to add additional resources within minutes by activating some of the

CSRs who have indicated via a Cloud-based work management system that they are available and willing to work. The CSRs use connectivity to CRM and KMS based in The Cloud, which gives them exactly the same capability as a CSR sitting in a contact center. Fourth, KMSs have become more comprehensive and easier to maintain, making information resources more effective. Finally, artificial intelligence systems have become sophisticated enough that they can draw on the KMS and respond effectively to between 20 and 60 percent of the email and chat questions presented. I have seen experiments in predictable scenarios, such as donation calls to a disaster relief organization, where voice recognition and voice response are so effective and lifelike that most customers surveyed afterward stated that they believed that the call was handled by a human being when, in fact, it was a computer.

Information Required

This part of the customer access plan should specify three types of information. First, what information is needed to describe the customer, the nature and the complexity of the issues, and therefore who can best handle the customer's request. This information must be captured by the access system so that the serviceperson can handle the issue when it arrives. Second, what information must be captured during the interactions? Third, what information must be captured to meet applicable privacy and reporting requirements?

Three big mistakes in information collection are to collect too little information to allow proper routing, to collect too much information, deterring the customer from continuing on to the service system, and not to pass on the collected data to the response process, such that the CSR has to, for example, ask for the frequent flyer's name and number again. A best practice is assuring that data need be collected only once. For example, a health insurance company uses voice recognition technology to create a record for future verification that a woman calling on behalf of her elderly mother is, in fact, authorized to talk about her mother's health care. This allows the company to comply with health information privacy requirements while eliminating the need to get the mother's permission each time to talk to her daughter.

Analysis and Improvement

Analysis and improvement focuses on how the access process itself can be continuously improved. This component includes gathering feedback and operating metrics from search, self-service, and IVR use as well as the automated call director (ACD).

A big mistake made by most organizations is not capturing information beyond how quickly calls are answered or how long the retail line is. Even the wait time metric is usually reduced to an average speed of answer. The beginning of this chapter stresses that you must first motivate customers to seek assistance, provide them with their preferred channel when they want to use it, and finally make using that channel effortless. Most companies do not collect information on the success or failure of each of these phases. I will address specific metrics in the next section.

Strategy for Introducing New Services

The access strategy must allow for the incorporation of the surge of assistance requests resulting from the introduction of new products and new marketing campaigns. Chapter 1 showed that up to 60 percent of service needs are caused by marketing, products, and policies. While many of these contacts can be avoided, every new marketing program or product introduction will create some surge of requests for information and assistance. In most companies, the access and service systems are forced to react to the new wave of contacts. A best practice is to have a rigorous strategy to flag in advance new products and campaigns that will impact contact volume and mix. If the access strategy has an overt process for identifying and planning for each new source of contacts, then fire drills, at least from internal causes, are reduced. This plan must have buy-in from the marketing and product development departments. One major pharmaceutical company required every new marketing and product management executive to attend a briefing on how to use the 800 number and the damage that could be caused if coordination is poor.

Metrics to Manage Accessibility

The primary metrics most organizations use to measure and manage access are wait time and abandoned service transaction. These are used for phone calls and IVR interactions as well as in retail environments. Unfortunately, these metrics are very blunt instruments. An average wait time of two minutes can include 80 percent of customers who are waiting less than 90 seconds and 5 percent of customers with wait times of more than eight minutes. Likewise, some customers will abandon their calls within 30 seconds, but most will wait over three minutes before abandoning. Therefore, even the rate of abandoned calls must be viewed in terms of the percentage of customers who have an unsatisfactory experience.

Metrics That Measure True Accessibility

Accessibility metrics should include at least four dimensions that stretch from the need for assistance to arrival at the appropriate response function:

1. *Noncontact rate*—At least yearly, the company should survey a random sample of customers to determine the percentage who have encountered significant questions or problems but who have either not bothered to make contact or have tried and have given up. Survey methods are discussed in Chapter 7.

2. *Percent waiting an unreasonable amount of time*—Unreasonable is in the eye of the beholder, but for almost all customers, waits of more than two minutes become irritating. Reports are needed on the percentage of customers who have to wait more than two minutes on the phone and more than four minutes in a retail line.

3. *Percent opting out of routing process*—Opting out includes both hitting 0 for the operator and abandoning the call after a failed transaction.

4. *Percent abandoning the contact attempt*—For all abandons, including from the IVR and FAQs on the website as well as the ACD, you should know both the volume and exactly where in the service process the customer encountered frustration or failure.

Technology can make much more data available if service and CE managers are willing to demand it from their IT departments and technology vendors. Your CRM, website hosting systems, IVR, and voice recognition systems all have data that describe the number of attempted transactions, the number of known transaction failures, and where in the transaction or service process the failure took place. I am not minimizing the challenge of getting this data in usable format, but if the data is available, it is a perfect map of your CE frustration points.

GETTING STARTED
Questions to Ask Yourself About Your Accessibility

1. Do you solicit contact whenever and wherever the customer is using or thinking about your product or service?

2. Is service available at all times, or, if not, do you at least set expectations on when a reply will be received?

3. Do you have a complete picture of your service workload at all customer touch points and the extent of service access failures?

4. Do your touch points have the training and resources to either handle the workload or get it to someone who can?

5. Is there an explicit process for revising the access plan when there are changes in products or marketing messages and/or in external events?

KEY TAKEAWAYS

- Your company must aggressively solicit requests for assistance because customers often feel it will do no good to make contact, it is too much trouble to contact, they will create conflict, or their preferred channel for contact is not available.

- Contact volume is driven by the volume of customer problems and questions, as well as by customers' involvement, the visibility of the communication channels, what each channel promises, and the availability of the channel, including wait time and welcoming.

- An access plan must be based on accurately forecasted workload by segment, which must then be translated into the resources needed, routing, service levels, and what service will be provided when and by whom and with what capabilities and, finally, by how new products will be accommodated.

- Metrics beyond average speed of answer and caller abandons must be extracted from the system technology to describe total transaction failures, the noncomplaint rate, the wait time distribution "tail" for the longest waiting 5 percent of customers, and the number and causes of abandons by communication channel.

Notes

1. Brad Cleveland, *Call Center Management on Fast Forward* (Colorado Springs, CO: International Customer Management Institute, 2012).
2. Scott Broetzmann, Marc Grainer, and John Goodman, *2013 National Rage Study* (Alexandria, VA: Customer Care Measurement & Consulting, 2013).
3. Full disclosure: I liked this concept enough to have invested in it.
4. Cleveland.
5. Cleveland.
6. Cleveland.

CHAPTER 6

Always Satisfy Customers, and Sometimes Dazzle Them

At one company I worked with, a finance department executive said he wanted to give good service but did not want to give away the store. When I asked why he thought a major giveaway would happen, he talked about the time Nordstrom took back and provided a refund for tires when they did not sell tires. Great service is not about giveaways; in fact, giveaways often do not work. Great service is about consistently delivering good service with an occasional flash of inexpensive delight and emotional connection.

Going from good to great service is often almost free. A chemical company delivered its products to customers via tank trucks driven by independent owners. The truckers had appointments but often missed them due to traffic. Missed appointments meant they had to wait in line, often in the hot Texas sun. Further, while the chemical was being piped into the truck's tank trailer, the company representative and the trucker would both stand and stare at the pipe for ten minutes without speaking. When I commented on the lack of interaction, a manager said: "They aren't customers—they just deliver to the customer." Interviews revealed that at the customer site, the truckers often made unfavorable comments about the company while they were unloading their chemicals. These noncustomer truckers were thus a key source of negative WOM. After discussions, the company provided a waiting area in the air-conditioned office and instructed the staff loading the tankers to initiate conversations with the truckers. Customer satisfaction improved at very little cost, just by providing a cool room and a little conversation.

Your customers expect you to be easy to do business with (ETDBW). ETDBW is a broad term that includes designing, evaluating, and upgrading service delivery across every customer touch point. You may need to break down outdated stereotypes of service and sales roles and responsibilities, and you may have to rethink mantras that are dead wrong (e.g., "The customer is always right"). Some of your low-value (as in bronze or lead) customers may be other company's gold customers—and could become your gold customers if you treated them better.

In this chapter, you will learn:

- Five required objectives for any service system and how those objectives make your company ETDBW.

- Six functions that will equip the service system to effectively achieve all five objectives.

- How technology can tailor the service system to provide the intensity of service desired by the customer.

- The key metrics needed to ensure that you are achieving all five objectives.

Five Objectives for Your Company's Service System

The primary objective of the service system is contact resolution of all customer requests, questions, and problems, either immediately or in a timely fashion. However, four other objectives can have as much or greater impact on the overall customer experience: creating an emotional connection, preventing future problems, up-selling and cross-selling, and input into the VOC process. All five objectives are discussed below.

First-Contact Resolution

First-contact resolution (FCR) is the traditional objective of all service systems. My research, going back to two studies performed with The Coca-Cola Company in the 1970s when rolling out their 800 number, has consistently shown that whenever customers contact a company a second time, satisfaction is reduced by at least 10 percent.

Emotional Connection

Emotional connection starts with the greeting, which establishes rapport with the customer. If a retail staff member is serving another customer, the greeting begins with a nod or verbal acknowledgment. Once the interaction formally begins, connection should be created in the ways described further in Chapter 9.

Problem Prevention

Prevention of both future problems and future contacts is the third objective of any service interaction. The serviceperson or response system should use the facts at hand to identify as many future questions or problems as possible and proactively offer advice or a caution to avoid any future issues.

Up-Selling or Cross-Selling

Only when the customer is fully satisfied with the basic reason for contact should the service rep try to cross-sell or up-sell. The only exception is if the additional product will explicitly help resolve the problem or prevent future occurrences of the issue. Then, the cross-selling action can actually become a source of customer delight. The decision whether to cross-sell or up-sell and, if so, what to offer should be data driven. The third section of this chapter will address how technology facilitates these activities. Many organizations make cross-selling the second most important objective, ahead of connection and prevention. But cross-selling without connection and preventive education seldom succeeds.

Input into the VOC Process

Less than a third of companies I have visited explicitly require front-line service employees to input information into the VOC process. My diagnosis of this failing is that executives either believe that be-cause it takes little time, the requirement need not be carved out as a separate responsibility or, worse, they have just never thought about it. The problem is that if input into VOC is not formally a part of the job description, the serviceperson will not feel the need to think about it, and supervisors will not feel compelled to evaluate it. The result is that very little input takes place in most companies—and if it does, it is ad hoc.

Easy to Do Business With (ETDBW)—Where Does It Fit In?

Being ETDBW is a major goal of most companies, but there is con-fusion as to what it means. My research has suggested that there are five components of ETDBW, many of which logically overlap other objectives of service but which go beyond them.[1] The key drivers of ETDBW are:

- *Ease of finding information and assistance*—Most customers go to a company's website or the welcome literature before re-questing service. Rarely do these resources provide adequate service information and thus are a major source of customer frustration. Many negative ratings in metrics such as the Cus-tomer Effort Score are caused by customer frustration in not being able to find where to get service or information.[2] It is ironic that this score results in the service system being blamed for a problem that often is not in its jurisdiction.

- *Ease of access of the service system*—This dimension was ad-dressed in Chapter 5 and is the most obvious indicator of not being ETDBW. Examples include frustrating phone trees, long waits on a phone queue or line, and truncated business hours.

- *Minimum of bureaucracy*—This dimension is one of the most insidious and hardest to fix. It requires minimizing the amount of information that the customer is required to provide and the number of people or steps involved in decision making. Both of these requirements are an anathema to risk-adverse organizations and their accompanying compliance and legal staffs. I will address this more in the response section.

- *Complete FCR*—Ideally with emotional connection and education for prevention.

- *Follow-through*—On promises made.

All five of the ETDBW drivers are represented in the CE framework. The two dimensions of availability of information and of the service system are parts of the access component, and the last three are contained in the service functions described in the next section.

Six Functions to Consistently Achieve All Five Objectives

Most service managers and contact-handling technology providers logically focus on the immediate tactical tasks at hand—get the contacts in, handled, and out. From a parochial perspective, this makes sense. From a strategic perspective, it is penny-wise and pound-foolish. Figure 6-1 introduces the framework for how to move from firefighting to anticipation and prevention.

The six key functions are a simplified distillation of the original framework of 19 functions in my first book, *Strategic Customer Service*.[3]

The six functions are logically ordered in the sequence of occurrence. Ideally, you want to anticipate service and then proactively deliver it. Anticipation is the first strategic function; many managers never even think about it. The second function is the ability to seamlessly intake contacts. Again, this function has significant strategic implications because it is here that you start collecting information on why the customer had to contact you. The why question reveals

Figure 6-1

Key Customer Service Functions

Key Service Functions
Anticipation of Needs
Intake of Request
Response
Follow-through
History
Evaluation

where future prevention and anticipation rest. The third function with strategic implication is response. Most companies define response as handling the contact to resolution. This is also when you have an opportunity to emotionally connect and prevent future contacts via education. The last three functions—follow-through, history, and evaluation—are primarily tactical if the first three are executed with an eye toward strategic activities.

Anticipation of Customer Needs

Anticipation uses existing knowledge in the Customer Relationship Management (CRM) system to identify and deliver a customer's future need before the customer asks for it. In some cases, this knowledge is in an operational database. If a communication company's database shows wireless calls dropping at the customer's home address, the company can suggest a new router or network extender—even if the customer has not complained. This proactivity converts an aggravation into a positive experience. Similarly, if a loan payment is not received by the day before the due date, the company can email or text the customer to confirm that the payment was mailed. If it has not been mailed, an online electronic or phone payment can be suggested to avoid a late charge. This too can be a delighter in that the company is valuing the customer's goodwill more than the revenue from the late fee.

Intake of Service Request

Intake of a service request has four steps. It includes the greeting, authenticating the customer's identity (for privacy and antifraud purposes), recording the issue, and classifying it. Ideally, doing this strikes a balance between gathering all the information needed to make decisions without gathering so much that the process becomes a burden to customers.

The greeting both confirms that the customer has reached the right company and sets the tone for the rest of the call. It can actually create emotional connection. For instance, at Frontier Communications, the CSR says, "Welcome to Frontier Communications. This is John Doe in Dallas. How may I help you?" Using both the first and last names and naming a nearby location all help personalize the greeting and create a genuine emotional connection.

Authentication usually follows the greeting, but if the customer has a simple question that does not involve account information, it may be an unnecessary aggravation. The website and phone center should have as much information outside the computer security firewall as possible to allow customers to obtain answers without going through authentication.

Recording consists of capturing all of the customer's personal information and the nature of their issue. Much of this can be automatically recorded from the automated number identification and linkages to the operations systems (e.g., so that the CSR knows what model product you have without having to ask).

Classification facilitates more rapid (sometimes automated) investigation of the issue to speed response time and also records information for several purposes. The information gathered during the classification step is used in the customer's history, in workload measurement, and for preventive analysis to eliminate future problems. For the contact to be useful for workload and preventive analysis, the classification must include the specific reason for the call (an overt problem such as "product does not function") and the general root cause (such as need for customer education versus a defective part).

Response

Response is the most complex function. There must be a complete response to all dimensions of the customer's request, accompanied with clear explanations, empathy (when appropriate), and possible emotional connection. Response consists of investigation, response formulation, connection, prevention, and transmittal.

Investigation is often the most labor-intensive because it consists of gathering information. In an insurance or technology environment, multiple databases may contain customer account details, the specifics of the customer's policy and their recent claim, and/or the particular technology configuration they own. In a service situation, most of the details, in terms of what the customer wants and relevant background, must be gathered directly from the customer. To avoid bureaucracy, limit data collection to what is critically needed, and which cannot be obtained internally. Photos uploaded from smartphones may replace long descriptions or the need for product returns.

The CSR's decision-making criteria and empowerment must be flexible enough to use the facts at hand to negotiate a solution with the customer. The response can be a one-time, one-way communication, such as, "I am sorry the hamburger is not rare as you ordered it, I will replace it" or, more often, an iterative process. If the suggested solution is not obvious or not exactly what the customer requested, the response should include the rationale for it, and if there has been a problem, a repeated apology that does not accept blame. Often, the response shows empathy and builds rapport.

Response formulation conveys the outcome of the investigation and the decision to the customer. This can be an answer to a question or, if there was a problem, a suggested solution or remedy. Customer Care Measurement & Consulting's *2013 National Rage Study* found that if the response includes both a tangible remedy, such as a refund or replacement product, and an emotional component such as an apology, the response is twice as likely to achieve high satisfaction.

Depending on the amount of time available, the severity of the situation, and the demeanor of the customer, the response may also include an attempt at emotional connection or customer education to prevent future issues.

The response is often transmitted verbally to the customer, but electronic responses are growing rapidly. Either way, other parts of the organization should be made aware of the issue. Transmittal of the response to the customer should always create a summary or copy of the response in the history section of the CRM system, which often provides the input data to the VOC process and notification to other relevant parts of the organization.

Follow-Through

Follow-through consists of the management and actual delivery on any extant or outstanding promised actions associated with the response. This includes actions that must be taken by another part of the company, such as adjusting an account, shipping a repair part, or making a service visit. There should also be an ongoing quality inspection function that samples responses to ensure that all of the actions are documented and that all aspects of the request were fulfilled.

Follow-through should always track the serviceperson or unit handling the initial contact (showing that the contact was handled to completion on first contact or that they had to call them back but did so as promised). This requirement for tracking of follow-through also applies to other partners within the company (e.g., field service staff who will make a repair at the customer's home) or even in another organization that delivers products or services. An innovation in this area is using the customer's response to the question, "How soon do you need this?" toward setting the follow-through time frame. I have found that most customers give reasonable time frames. This approach also allows resetting expectations if the customer appears to be unreasonable or for balancing workload if the customer is not in a hurry for a final response.

History

The entire interaction with the customer should be recorded in a manner that facilitates two future activities. First, enough information must be collected to allow other employees to know what was done for that customer and what was promised. Second, the history must have enough detail for analysts to understand the general root cause of the contact and to facilitate VOC analysis.

Many organizations do not record simple contacts because they think the labor expended to record the contact is greater than the labor needed to handle it. If the CRM technology is simple, incremental labor to record the contact is low or nil, and if you do not record the contact, you cannot identify a way to quantify or prevent future similar problems. Second, if the customer called with a simple question, this is an opportunity to make a connection or do prevention. Finally, if the contact is not recorded, the customer history is incomplete, which may hinder the understanding of more serious future problems. For example, a customer's anger with a delayed service call might be more understandable if a CSR is aware that the customer had previously called twice and had been told that the technician was scheduled to arrive before noon. Such confirmation calls are often not logged because at the time they seem trivial.

Evaluation

Evaluation consists of measuring the timeliness and effectiveness of the response process. There are three methods. First, customers can be surveyed about the content and timeliness of the service. Surveys are an easy way to measure customer satisfaction with content. Measuring the customer's perception of timeliness by means of a survey is more difficult and imprecise because customers often have misperceptions of how long they were on hold or how long it took to get a final response.

Think how long a minute seems when a service representative puts you on hold; a minute on hold seems much longer than a minute. My clients have conducted experiments showing that customers' perceived time on hold is often two or three times longer than the actual time. Also, if the survey is conducted using an IVR survey tool immediately after the call, the customer cannot comment on final follow-through because it has not happened yet.

The second approach is to evaluate the actual contact record and recording. A quality analyst can listen to the call, read the emails, and review entries in the record and history.

The third approach is to analyze the contact using speech or text analytics. In this approach, the flow of conversation is searched for key words and phrases, as well as for tone and loudness of voice to identify emotions.

Regardless of the method used, the outcome of the analysis should be used for evaluating both the serviceperson and the response process for that type of issue. Usually when the customer was not happy, the CSR followed the response protocols but the standard response was not successful in satisfying the customer. Thus, the protocol should be reviewed and the CSR not penalized.

Applying Technology to Create High-Tech High Touch

The following are the primary technologies that should be used to move each set of functions to the next level.

Anticipation

There are two types of anticipation actions. The first prepares the CSR to effectively deal with customers by providing all the critical information to fully anticipate their needs and sensitivities—their history of contacts, survey results, preferences, and hot buttons. Companies also may use information about the customer's experience to proactively reach out before the customer contacts the company. Both actions can have significant impact.

The CRM system will contain the customer's history and preferences, the results of customer surveys, and employee comments. Behavioral modeling software tools use demographic and psychographic data to predict behavior and preferences. These tools can also scrape all existing data from social media and combine it with survey data to create models that predict the hot-button issues that will motivate customers to act. For example, a model can predict what someone will buy or how they will vote. Further, these results can be used by CSRs to guide the approach to each interaction and to create stronger customer connection and perceived product value. For instance, an auto customer who strongly supports green initiatives may be interested in buying a high-mileage car and learning about fuel conservation. Several tools providing this kind of information are on the market now; one that I am familiar with is Resonate, which supports digital marketing programs.

Proactive service is a more activist approach. Here, the company mines operational data to flag unpleasant surprises and reaches out to the customer before the customer contacts the company requesting service. Such technology now incorporates sensors in aircraft, copiers, computers, and even home security systems that transmit product performance back to the company. Such wireless/mobile linkages allow companies to go beyond identifying product failure (such as the office copier that is overheating). They can also tell companies when a customer is not using the full range of the product's functionality, allowing the company to educate the customer. For example, the customer could be educated on how to use additional aspects of the car's navigation or entertainment systems that they are not currently using. The more functionality customers use, the more value they perceive.

The actual education can be through phone calls tailored to the customer's schedule, an online chat, or a series of short 45-second videos. (Five 45-second videos have a much greater probability of being viewed than one four-minute video, given most people's attention span.) As mentioned, Tesla provides a welcome call prior to delivery as well as 28 minutes of video that are explicitly broken into 16 segments, each 1–3 minutes in length. Customers rave about the quality of the delivery and education process.

Implementing such anticipation actions based on operation reports can be a challenge in many companies because operations data systems are not usually designed to report on incidents that can be linked to particular customers. You should experiment with obtaining data on one or two of the most prevalent unpleasant surprises or educational opportunities provided to customers and use that data to anticipate the issue and demonstrate the positive payoff. Southern California Edison, a West Coast utility, chose to help customers anticipate their high bill. Smart meters allowed the utility to identify increase in consumption and warn customers of an unanticipated high bill—one of the worst surprises in the power industry, other than losing your power.

Intake

The automated call director and IVR, both discussed in Chapter 5, are the first phase of intake for telephone calls. In many cases, customer identification can be verified with automated number identification (ANI). A second easy approach is to use voiceprint technology, although biometrics is still in its infancy for private sector applications. In retail environments, customers making returns are asked to provide receipts. Although many do have their receipts, this requirement creates an additional barrier to reporting problems. There is a trade-off between the inconvenience of the barrier and the cost of fraud from dishonest customers.

The website, as connected to the CRM system (possibly assisted by speech or text analytics), is responsible for the bulk of the intake, such as recording the reason for a call, history, and classification of the issue. The problem with gathering the necessary detail is the time needed to accurately classify contacts without creating undue bureaucracies. Two technological solutions now exist. First, if the call is billing related, the CSR notes the code for billing, and the CRM system provides nine subcategories from which to choose, thereby eliminating the need to look for the subcodes. This automated, hierarchical coding approach very efficiently provides detailed coding. The second approach is to use speech or text analytics to do the coding.

Speech analytics can identify and code three key attributes of calls: subject, emotion, and outcome. The subject of the call can include what product the customer has and the question or problem. The emotion can identify irritation, anger, or happiness. The outcome can be the customer's relative satisfaction with the overall response. Two limitations to speech analytics still exist. First, the software must be carefully tuned to ensure accuracy. Second, it still has difficulty identifying the cause of the caller's problem.

Finally, as many consumers are reluctant to fill out forms online, some organizations have humans facilitate the process. As noted earlier, Philadelphia's Philly311 service center uses volunteers who publicize the service by going to local meetings, telling their neighbors about it, and offering to enter complaints and service requests on their behalf via the website. They are also trained to set proper expectations, such as explaining the city's process for getting potholes repaired.

The intake process will also negotiate a time frame for response (ideally by asking customers how soon they need or desire an answer). It will also gather the facts needed, such as a claim number or production code, in order to begin investigation during the response phase.

Mobile technology is complicating intake because standard laptop or desktop Web screens are not easily directly translated to the smartphone screen. Input has been further aggravated by the migration from pushbuttons to touch screens that have reduced accuracy. On the other hand, the cameras on smartphones have allowed insurance companies to conduct preliminary analysis of accident damage accurately enough to authorize repair and be relatively confident of correct assessments.

Response

All five of the response activities are very dependent on CRM, Knowledge Management Systems (KMSs), and operational databases. The most dependent activity is investigation of the issue prior to formulation of the response. The CRM identifies the customer's products and history. The KMS suggests the flexible solution spaces and general response rules. The operational databases, when linked to the CRM automatically, indicate the actual status of the customer's particular transaction, greatly reducing bureaucracy.

The biggest problem most organizations encounter in supporting a response system is keeping the KMS up-to-date. Even small service systems must have a full-time KMS administrator who systematically canvasses the rest of the organization for changes in products, marketing offerings, and service policies. The good news is that the best KMS programs automatically administer the response rule and knowledge update process.

A completely different approach to response support is the online peer-to-peer community or forum, which is hosted either by the company or by an independent group of customers who all use the same products such as Apple Mac users, Chick-fil-A customers, or Auto-Trader customers. Customer experts who share their expertise and help in communities are not paid by the company but might get special access or recognition if they become power users or prolific supporters. These incentives are known as gamification strategies. Companies such as Intuit have used them to encourage accountants who use its software to provide answers to their peers. A majority of issues are answered by the community, deflecting workload from the Intuit support team. Another example is Stack Overflow, a collaborative platform where technologists help each other address difficult technical challenges. On their website, community members post technical dilemmas and get help from other members who are rewarded with reputation points as determined by other members and the person seeking help. Communities must be hosted and monitored for inappropriate behavior just like a company's Facebook page, but in general they are very cost-effective. I will address the operational challenges of communities in Chapter 8.

In summary, technology makes response more tailored and efficient. Response formulation primarily takes data from customer history, which is stored in the CRM, and combines it with the recent customer request (from Intake) and makes a decision using the rules and parameters in the KMS. Response formulation can also use speech or text analytics or the online community to manage the investigation and tailor formulation of the response with the customer. Speech and text analytics can be used in real time to classify communications according to the reason and the root cause of the contact and also to ascertain the level of customer satisfaction with the response. This assessment has been used for real-time analysis to identify customers who are getting upset at the moment and notify a supervisor or coach to tap into the conversation to determine whether to intervene on the spot.

Additional aspects of response formulation include creating the emotional connection and preventive education. At Zappos, CSRs are trained to actively search for opportunities to make a personal emotional connection. CSRs listen for information that falls outside the context of the transaction and tactfully delve into the subject matter to extend the range of the conversation.

Follow-Through

Follow-through ensures the complete and timely delivery of the response. In organizations with high first-contact resolution, most follow-through is instantaneous. When the first contact does not resolve the problem, the CSR uses the CRM system to instruct operational systems to execute the request. If the operational back office electronically confirms receipt of the request and execution of action, the CSR is able to say confidently, "It has been done!"

Follow-through becomes much more problematic when the CSR requires human action—when a part must be shipped or a system must be installed. Because these actions are taken by other units, follow-through must be supported by service-level agreements (SLAs) between units promising to take action within a certain period of time. The CRM should inform the CSR about how reliable the other unit is, so the CSR can give the customer accurate information—and do so with confidence. When the CSR projects this confidence, customer follow-up calls asking whether the transaction has actually been executed fall by more than 90 percent.

History

The history should reside completely in the CRM system. This can happen only if all other systems, such as operational systems, include a customer identifier for each transaction. Whenever a customer's account is called up for a service interaction, effective CRM systems show the CSR the last three interactions the customer had with the company.

Evaluation

The CRM system must contain all the data needed to execute the customer surveys: preferred channels for contact, the most recent transaction, and positive or negative feedback. The CRM system must also have the embedded tools to execute such surveys via multiple channels including immediate email surveys. Immediate surveys should *not* be dispatched if promised actions are still outstanding. It makes no sense to survey customers waiting for actions to be taken. Many organizations do immediate after-contact surveys because the surveys are used to evaluate the staff. I disagree with this approach for transactions that require further action because the most important aspect of the transaction, the final resolution, has yet to happen. Thus, survey methods should vary according to the transaction.

Quality monitoring tools can record conversations and capture both speech and all keyboard entries to the CRM, or they can simply analyze the recording for key words, phrases, and overall satisfaction and emotion. Such technology is still imperfect. However, a number of companies have found it effective enough to eliminate the majority of their call evaluation staff, letting the software sort through thousands of calls to identify the particularly good and bad calls for human evaluators to review and use for personal coaching of the CSRs. Within two years, speech analytics will replace 50 percent of all call quality staff.

The final evaluation tool is the ACD, which captures talk time and after-call work, usually called wrap time, and links it to the specific CSR. The best evaluation processes link both back to the call record in the CRM system. This allows analysis of talk time by type of call and identification of best practices by type of call.

Metrics to Manage Service

There should be metrics to allow the evaluation of both productivity and effectiveness for each of the five objectives: first-contact response, emotional connection, prevention, cross-selling, and input into the VOC.

Metrics to Support All Five Objectives

This section outlines the key metrics for success and the pitfalls of several commonly used metrics.

First-Contact Resolution (FCR), Satisfaction, and Willingness to Recommend

FCR ratings should be triangulated by asking the CSR and the supervisor some of the same questions customers are asked in the satisfaction survey. First, was the issue fully handled on first contact? Second, was the customer satisfied? The satisfaction question should use the same scale or structure as the customer survey. Finally, how likely is the customer to recommend the company? Many companies ask the first and third questions without asking the CSRs whether the customer was satisfied.

FCR and Satisfaction by Type of Reason for Contact

The results of the FCR and satisfaction questions should be analyzed by the reason for contact and, if appropriate, by product type. This analysis will highlight the issues and products for which the response rule or solution space does not consistently achieve high levels of resolution and satisfaction. In such cases, the problem lies not with individual CSR responses but with the formulation process, the CSR response guidance, or the KMS content being used.

Follow-Through

Follow-through metrics ensure that the basic response process and FCR are completed. There are two metrics for measuring follow-through. The first is the percentage of contacts that are handled within the response time standard (e.g., 90 percent of emails are answered within four hours). The second is the percentage of transactions assigned to another responding unit or partner (such as dealers or local franchisees) that were not handled within the time and quality standards. This second metric is more critical because failure undermines CSR confidence in the other responding units. A lack of confidence creeps into the language used by CSRs when talking with customers. This will result in unnecessary callbacks to confirm that promised actions were performed.

Connection and Prevention

This metric asks whether the CSR created an emotional connection and/or educated the customer on how to avoid future problems. In many cases, education is actually a delighter and prevents future workload. This metric should be collected as the percentage of contacts where the CSR intentionally went beyond the basic subject of the transaction to create an emotional connection via preventive education or simply a discussion about the customer's location or pet. Once these contacts are flagged, it will then be easy to measure their impact by including those contacts in the sample for satisfaction surveys.

Cross-Selling or Up-selling

Cross-selling and up-selling measure the number of opportunities identified by the CSR and the degree to which they are converted into sales. The usual metrics for this are the number of CSR attempts and sales successes. The attempt metric can be self-reported but is open to manipulation. The key metric most service systems use is actual sales.

Input into VOC

This metric creates accountability for identifying opportunities for problem prevention, enhanced CE, or improved processes and for making input to the VOC process. Ideally, the root cause analyst in the contact center keeps track of the number of useful VOC messages received from each CSR. There should be a minimum number per quarter, and performance recognition should be given to the CSRs who provided the most useful input. Contact and service staff should all receive a monthly email describing some of the main actions taken based on their input to demonstrate that it is actually being used.

Metrics That Should Be Used Carefully

The following four metrics are useful but often overemphasized or misused in attempts to simplify management of the contact center or service system. They are all averages or summaries of much more complex underlying sets of data, which must be understood if the proper decisions are to be made.

1. *Average speed of answer*—Average speed of answer (ASA) can be extended up to 60–90 seconds with little to no damage, if the call is effectively handled.

2. *Average handle time*—Average handle time can penalize CSRs who happen to encounter more difficult calls, forcing them to truncate service on the calls that make the greatest impression on customers.

3. *Net Promoter Score (NPS)*—This metric is a blunt instrument that ignores customers who give a score of 7 or 8 out of a possible 10 (called passives in the NPS process), which could include 30–40 percent of all customers.

4. *Customer Effort metric*—This metric penalizes contact centers and CSRs by including the customer's difficulty in finding answers on the website or who or where to call. This is usually the website's fault, not the CSR's.

GETTING STARTED

Questions to Ask Yourself About Your Service Process

1. Do you have prevention and emotional connection as formal objectives of your service system?

2. Have you given your front line the tools needed to be successful?

3. Are you measuring service success by type of issue?

4. Is your website an effective first stop for answers and problem prevention in your service system?

5. Is your quality monitoring system measuring what truly drives customer satisfaction?

KEY TAKEAWAYS

• Five objectives that should exist for any service system are first-contact resolution, connection, prevention, cross-selling, and input into the VOC process.

• Six functions that will equip the service system to effectively achieve all five objectives are anticipation, intake, response, follow-through, history, and evaluation.

• Technology should tailor the service system to provide the desired service via the customer's preferred channel; this channel will often change during the transaction.

• Performance metrics should be used to ensure that your system is achieving all five of the service objectives.

Notes

1. John Goodman, "Nice Doing Business with You," *iSixSigma Magazine*, January/February 2011, pp. 43–46.

2. Matthew Dixon, Karen Freeman, and Nicholas Toman, "Stop Trying to Delight Your Customers," *Harvard Business Review*, July 2010.

3. John Goodman, *Strategic Customer Service* (New York: AMACOM, 2009).

CHAPTER 7

Listening Passionately to Your Customers' Unified Voice

A major communications company was spending over $15 million annually surveying customers and compiling data. The joke was, "If it moves, we measure it." However, as the vice president in charge put it, all they did with the information was "review it each quarter to see if things had gotten better or worse." He added that because even those quarterly reviews were poorly attended, half of them were cancelled.

Contrast this company to one of its competitors that spends less than $1 million annually on VOC. This company harnesses speech analytics technology to analyze and report on conversations with customers in real time. It draws on multiple sources of data, including operations data describing outages and field service visit failures, as well as surveys. It then uses technology to integrate the reports and interpret the data to create actionable information used by every corporate function to improve the CE. The company's robust employee input process enables employees to report frustrating interactions and opportunities for improvement of products, processes, and policies into the same VOC system. This company spends less on the VOC process than many other organizations do, but it gains dramatically more impact from technology to provide a complete view of the CE by combining surveys, customer contacts, customer preferences, operational data, and transactions.

In this chapter, you will gain a more detailed understanding of how to create an effective VOC program or improve an existing one. Specifically, the chapter explains:

- Appropriate objectives for a VOC process.

- The eight building blocks for a comprehensive VOC program.

- How to use both mobile and social media inputs.

- Four success factors that ensure VOC process impact.

- Common implementation challenges and how to avoid them.

- Practical tips for enhancing your existing VOC's effectiveness and impact.

VOC Objectives

VOC processes can have different objectives for different company functional managers. To design engineers and quality professionals, VOC helps determine the requirements for creating the ideal product for the customer. To marketers, the VOC process uses survey and research tools to understand how to attract new customers and enhance sales. To customer service executives, the VOC process provides feedback on the tactical success of the service process. But for executives concerned with the overall CE, the VOC must describe the entire, end-to-end CE—not just the quality, marketing, or customer service component.

A VOC process should help companies:

- Identify emerging issues.

- Set priorities for addressing new opportunities to improve the CE.

- Track progress on previously identified, ongoing problems.

- Create the economic imperative for action on the CE by the whole organization.

- Assign ownership of CE improvement projects.

- Quantify in a credible manner the financial impact of CE improvements.

VOC data should also be available in a timely and detailed enough manner to:

- Prevent problems by understanding their cause (employee, product design, marketing, delivery, or customer use?).

- Intervene, when it makes sense, while the problems are happening.

- Improve products and processes over the intermediate and longer term.

One final caution: VOC refers to systematic information on the experience of current customers. A separate sector of market research focuses on winning new customers. Yet many managers and professionals view market research surveys as the mainstay of the VOC program or as the program itself. This is mistaken for three reasons.

First, the majority of market research is aimed at winning new customers. Therefore, combining market research and VOC confuses the study of existing customers with that of the (usually much larger) noncustomer market.

Second, surveys taken days, weeks, or months after a purchase or experience are lagging indicators. An electronics firm I worked with discounted surveys, rationalizing that they had already moved on to making redesigned models of the product.

Finally, surveys about prospective or hypothetical products and services can be quite unreliable. A classic historical example occurred in the late 1970s. Customers overwhelmingly rejected the idea of automated teller machines (ATMs), saying they would not trust machines with their money and wanted to interact with real people. Yet once they had round-the-clock access to cash, they quickly became fans. We have seen a similar progression with online check-in for airlines.

Although surveys and focus groups provide important information on the CE, they provide relatively little data on the internal causes and operational details of customer experiences. What is needed is a panoramic view of the CE provided by data from a myriad of sources.

Key Building Blocks of an Effective VOC Process

Figure 7-1 shows the focus and sources of data for an effective VOC process, which must include input from all phases of the CE. It is far more than feedback from contact management (what most people call customer service). The VOC process also includes input based on the customer interaction with the product design, marketing, sales, product delivery activities (DIRFT), and the access strategy activities, as well as the service activity. The VOC process should also include descriptors of the CE that do not come directly from the customer. These can include operations that will adversely affect the customer— for example, missed deliveries, bounced checks that will produce a service charge, and higher water utilization that may mean a leak in the water supply.

Figure 7-1

Key Components of a VOC Process

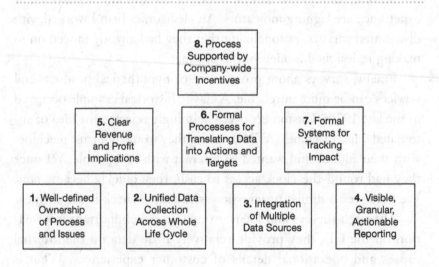

As noted in Chapter 3, few organizations have effective VOC programs. A recent study I led found that only a third of VOC processes fixed the majority of issues raised by customer feedback.[1] The four attributes most associated with an effective VOC process with impact were:

1. The VOC process gathered data on the end-to-end CE, from expectations setting to final product use, ensuring that all phases of the CE were examined.

2. The VOC process drew upon and integrated multiple sources of data to enhance the credibility and impact of the data, findings, and recommendations. Sources should include surveys, complaints, customer contacts (both service and sales), social media, operational metrics describing what the company has done to, or for, the customer (such as warranty claims, missed appointments, invoice adjustments, late charges, and missed shipment dates), and employee input.

3. A unified view means a single agreed-on reality based on multiple data sources describing all the phases of the customer lifecycle.

4. The finance department and CFO's buy-in and acceptance of the validity of VOC analysis and output were crucial to VOC having an impact and getting the majority of issues fixed.

Also, the study found that 20 percent of the 160-plus responding executives *did not even track* the percentage of issues identified by the VOC process that were resolved. You cannot have an effective VOC process if you do not track the percentage of issues identified that get fixed. Therefore, an effective VOC process must provide an end-to-end, unified picture of the CE and its financial implications that the CFO accepts.

Unfortunately, data input to most VOC programs does not fit together! Given a choice, each department will collect the data and use the formats that are easiest and most useful for its functional area. I have found as many as seven owners of different data sources feeding the VOC program, resulting in fragmentation and inconsistency in data collection and analysis. However, this messy organizational issue can be overcome if the following eight success factors are accepted and implemented in the company. They are all logical and are usually accepted and endorsed by management.

Having One Executive in Charge

Putting one executive in charge of the overall VOC effort, or at least the coordination of it, will reduce fragmentation and provide guidance on how everyone collects data in their area of responsibility, thereby ensuring VOC data compatibility. This leadership role can vary from a limited part-time voluntary VOC coordination role up to the senior executive role of the CE leader. Chapter 10 discusses possible roles of such an executive position. When this position or role does not exist, the head of quality or service, the chief operating officer, and the chief marketing officer are all candidates.

Unified Data Collection Plan

Not all VOC data must be collected by one functional area or unit in a company; that would be impossible in a large organization. The data collection must be guided by a plan that allows the data to be pieced together and reconciled using a classification scheme that is unified or at least coordinated across all silos. The VOC manager must create the scheme and enforce the plan.

A unified data collection plan ensures that all departments describe customer problems the same way in order to validly quantify and understand the problems. At an automobile manufacturer, the factory engineers, sales and marketing functions, customer service, and the dealer service technicians all had different methods of describing the same customer problems. The engineers talked about subassembly failures and customer maintenance failures, while the service writers and service technicians spoke about symptoms such as brake pulsation and engine hesitation.

In contrast, a number of years ago, Ford Motor Company began to collect consistent descriptions of problems and symptoms. One aspect of this plan recognized the importance of customer descriptions of problems to service writers when cars were submitted to dealers for repair. They created a small brochure to assist customers in describing the symptoms of their problem with the whimsical title, "Is it a Rattle or a Click?" This helped customers use the right terminology and ensured the correct diagnosis and repair.

The data collection plan must also use new data sources such as social media. Traditionally, all data input to a VOC process consisted of problem codes and technical entries by service writers, CSRs, and service technicians. There is now a flood of information describing problems and product behaviors and experiences from new sources, such as online reviews and social media. This data is textual in nature and traditionally would not even be considered data. Yet screen scraping and text/voice analytics can transform this free-form flow of customer and independent reviewer input into easy-to-analyze data using the same classification scheme.

Data Integrated into a Unified Picture of the CE

Data from all sources and channels must be combined and reconciled to create an end-to-end unified picture of customers' unfulfilled expectations and of the sources and impact of disappointment. Here are some of the challenges:

- The data from surveys, complaints, social media, and operations must all be compatible in terms of categorization of expectations, problems, causes, and impact on loyalty.

- Customer contact and operations data must be extrapolated to the entire marketplace in order to be comparable with the survey data, which tends to describe the marketplace as a whole.

An additional challenge is that VOC data must be structured to be used for multiple purposes across different parts of the organization. Common purposes include identifying emerging issues, intervening on the spot to restore customer satisfaction, taking action to prevent or mitigate current issues, and tracking progress on previously known issues. Further, the analysis should highlight the highest-priority issues based on actual or potential market damage.

To do a job effectively, people need the right information at the right time. Given employees' workloads and the pace of organizational life, this means proactively distributing not only data but also actionable analysis that connects the dots for the operational managers who are in a position to take rapid action if informed of the problem or opportunity. Interested parties should be able to access data, proactively push the information, and connect the dots for others, especially those who may need the data in order to take action. One of the more depressing occurrences at any company is learning that a problem could have been avoided if one department or group had been given information that another group already possessed. Sharing data with all who need it is a huge opportunity for technology. In the future, technology's role in industry could be as crucial as its role in national security.

Issues Must Be Actionable

However, it is neither helpful nor necessary to distribute all data to everyone. Data overload usually creates more problems than it solves. Those in charge of the VOC program must know who needs what, take the risk of filtering and tailoring the information to their customers' goals and needs, and proactively distribute a summary (ideally in at least two mediums, written and in person or video). The description of issues must be granular enough to be actionable. For example, "billing issue" is not granular; "incorrect late charge on premium accounts" is granular. The top issues, along with the monthly cost of inaction and a recommended course of action, presented in a one- or two-page summary works best.

Chip Horner, Global Head of Consumer Affairs at Colgate Palmolive, says he always follows up on the report with a face-to-face meeting. He wants to ensure that the information is understood and accepted and any questions get answered. Although tailoring reports to each function and reinforcing them via meetings is more labor intensive, the effort leads to higher impact.

Create an Economic Imperative for Action

Unmet customer expectations and poor service experiences have an impact on revenue, WOM, risk, and profits. Quantifying the monthly bottom-line impact of these issues (and the monthly cost of the status quo) transforms inert data into actionable information, creating an economic imperative. Unless the economic imperative to act is highlighted on the first page, VOC is just a feel-good exercise. Making the cost of inaction obvious compels action!

Another major weakness in the presentation of VOC economic analysis is that most analysis simply outlines the out-of-pocket company cost of customer problems, such as warranty service expense or the immediate lost sale. This implies that if complaints could be reduced, costs would decrease, and the bottom line would improve. In fact, the revenue damage is probably 10–20 times greater than the potential cost savings.

Consider two examples of the real cost of problems. When finance managers look at retail out-of-stock complaints, they think the lost sale is offset by the saved cost of carrying the inventory. In fact, when customers cannot buy what they came for, they are less likely to buy other items or return for anything in the future. Likewise, if they cannot get replacement parts quickly, customers are less likely to repurchase the brand. Low inventory has a significant negative impact on future revenue.

In a second example, progressive-thinking CFOs view an investment in improved quality and complaint handling as protection against regulatory action. One medical device company spent $100 million responding to a Food and Drug Administration warning letter. Their executives now view an extra $5 million annual investment in enhanced quality and patient relations as an insurance premium to prevent receipt of a future warning letter that could damage the brand.

Define Targets for Improvement and Suggest Potential Action Plans

Once an issue has been identified and the cost of not acting has been calculated, there are two other common impediments to fixing customer problems. First, the managers assigned to the issues must often start from scratch while continuing to do their regular duties, so fixing the problem is a collateral task. Second, they are rarely told what a successful outcome would be (total resolution is usually not possible). Suggested actions and achievable targets must be part of the VOC process. Some VOC analysts fear violating the prerogatives of line managers, but most managers are thankful for suggested starting points, action plans, and achievable targets.

On the subject of setting targets, one of the silliest exercises I've seen is so-called satisfaction planning for the upcoming year. A company that has, for example, a 76 percent loyalty level will select a target level of 80 percent for the next year. Why 80 percent? The rationale usually is that 80 is higher than 76. This is irrational target setting. A better alternative is to use the Market at Risk analysis (from Chapter 3), as well as an analysis of key drivers of satisfaction (to be discussed), to identify the high ROI improvements, assign responsibility for implementation, and use this information to suggest the expected increase in loyalty and satisfaction. Such analysis equips management with actionable plans for hitting satisfaction goals.

Track Results

Once actions linked to improving both the CE and the organization's financial performance are assigned, identified issues need to actually get fixed. Most companies file an action plan and then consider the matter closed. Rarely does anyone go back to check whether the action taken actually moved the needle. Effective organizations require process tracking and measure the percentage of issues raised by VOC that were actually addressed. Key questions are: Did the issue get fixed? For ongoing issues, what is the status of the proposed action plan? Are we making progress?

tions due to misleading marketing. Service quality metrics can include customer wait time on hold as well as evaluations of call content and how interactions were handled.

The big opportunity comes from evaluating the service delivery calls but with a different objective. Most evaluators listen to calls to determine whether the CSR followed protocol and was courteous and possibly to ask whether there is a better way of responding to the issue. Evaluators seldom try to understand the underlying reason for the call. There is great opportunity in asking, "Was this call avoidable?"

Complaint Data and Customer Inquiries

Many companies triage all customer contacts into complaints and inquiries. Complaints are usually coded separately and given more weight than inquiries. Often complaints and inquiries have the same content. Complaints are often just slightly more forcefully presented. On the other hand, questions by customers are rarely recognized as complaints. For example, when customers question a service charge, is that an inquiry or a complaint? If they say they cannot find a product and ask what retail location has it in stock, is the communication actually a question of where to buy the item, or is it a complaint that the product cannot be found? A further subset of these contacts is escalated complaints, such as when someone asks to speak to a supervisor or calls the office of the president.

Customer contact data is often produced in real time from customer interactions with the service system—mainly by phone, IVR systems, and email. This data has the advantage of being timelier than surveys and often has palpable emotional content. Recordings of customer complaints, played when presenting statistics to management or a manufacturing plant floor staff, have far more impact than numbers. Unstructured data, whether recordings or CSR case notes, can be converted into coded descriptive data either manually or with speech/text analytic software.

Contact data are harder to interpret because they represent only a small percentage of customers with questions. Most questions go unasked. Some issues (whether defined as questions or problems) are less critical (e.g., late charges) and less likely to result in calls than other issues such as missing parts for a Christmas toy. The multiplier for contact data, first introduced in Chapter 2, can be as low as 4 to 1 (implying a 25 percent contact rate) and as high as 100 to 1 or even 2,000 to 1. Only a percentage of customers will communicate on a problem or question, thus it is important to extrapolate contact data to the customer base and the marketplace so that it can be tied to survey and operational data.

Escalated complaint data includes not only what the original problem was but a reason for its escalation. The cause of escalation indicates where the primary response process is inconsistent or flawed. Customer contact data from all sources—call centers, Internet, and technical self-service—provide timely and robust input on the CE.

In addition, contact data can provide valuable clues about the causes of problems, including customer expectations and actions. All of these can help you prevent problems from recurring. For instance, a packaged foods company received complaints about mold in spaghetti sauce shortly after it eliminated preservatives for marketing and health reasons. Analysis of discussions with consumers revealed that complaints increased when consumers left the opened jar of sauce in the fridge for over two weeks. Simply adding a note to the label, "May be refrigerated for up to seven days after opening," dramatically reduced complaints.

Finally, positive comments can be used as staff motivators. A new mobile tool, 45sec.com, enables consumers to dictate up to 45 seconds of feedback to management on great staff performance (as well as complaints) with one touch of their mobile phone. The tool is location-based, and Google Maps and background tools do the work of finding the owners' email address and deliver the message to management within two minutes. Positive feedback can immediately be given to employees and celebrated in daily stand-up meetings, which is a huge motivator.

Mobile Contacts and Interactions

Every manager I know is concerned (some are freaking out, to put it bluntly) about the exponential increase in mobile interactions—but mobile is just another conduit. As Matt Trifiro, CMO at Heroku, a cloud application platform company, said, "Tweeting is the new 800 number."[2] Yes, you need the capacity to handle mobile workload, but it is like phone, email, and Web interactions. They are all part of your existing workload; they just come with higher expectations of immediate response and reaction.

Survey Data

Surveys can include complex relationship surveys, follow-up questions after contacts, and surveys while visiting websites. Surveys are the most reliable source of data on loyalty and on WOM impact. Surveys can also provide a view of the end-to-end CE from the customer's perspective. If based on statistically sound samples, surveys are more representative of the CE of the entire customer base than complaint and contact data.

Survey data does have some limitations. Its drawbacks include:

- The potential for bias created by survey design and sampling techniques. For example, phone surveys to landlines miss many twenty-somethings who have only cell phones.

- The cost per completed survey can be high when conducted face-to-face or by telephone.

- Most surveys require time to collate and process the data, whereas surveys associated with many CRM systems now produce results almost in real time.

- Surveys cannot identify why something occurred within company processes.

- Surveys are not very effective at measuring hypotheticals.

On balance, surveys are an important component of the VOC. But understanding the VOC requires more than just customer surveys.

Social Media, Online Reviews, Communities, and Other Unstructured Data

This information extends from the actual text of social media and online community interactions to online reviews and emails. When manually reviewed or, more practically, analyzed using text/speech analytics software, this information can provide valuable insight into a broad range of issues. Examples include the frequency and causes of problems, how often certain words are used, customer expectations, levels of emotion, the effectiveness of company responses, and damage to loyalty.

Social media and call recordings are also invaluable in humanizing the data. To say "85 customers had a problem" is interesting, but to attach an emotional quote dramatically enhances the impact. A New York utility used a report entitled "The Mood of the Mail," which included customer comments complete with swear words and sarcasm. It was one of the most widely read reports in the company.

As I noted in Chapter 3, contrary to popular perception, most consumers do not turn first to social media. In general, American consumers want to complain in private, which is where mobile apps like 45sec.com come in handy. A 2011 study by Intelliresponse found that fewer than 1 percent of consumer interactions with companies via social media were complaints.[3] This number matched exactly the finding of CCMC's *2011 National Rage Study*.[4] The *2013 National Rage Study* found that this number rose only to 4 percent, even for the consumer's most serious problem of the year, so social media is *not* a major channel for complaints, even in 2013.[5] On the other hand, a 2012 study by Argentina-based Proaxion found that 25 percent of Argentinians complained first via social media.[6] Suffice it to say, in this fast evolving space, no consistent rule can be applied.

Online communities, especially among business users, are good sources of intelligence on customer problems and attitudes. Intuit, the maker of popular finance and accounting software programs, sponsors an online community of thousands of accounting professionals. These users readily report problems and advise each other.

In many cases, Intuit's customers are more comfortable telling others in the network about issues, rather than complain to the company. Intuit facilitates forums and can observe and learn from the conversations. While many companies get most of their input from Facebook pages, General Motors gets only 7 percent of its social input from that source. The vast majority of social input comes from online communities hosted by car enthusiasts.[7]

Likewise, online review sites such as Open Table and TripAdvisor are great sources of diagnostic information about the CE, even though bogus reviews—both positive and negative—are on the rise. This is another reason for using social media as just one of multiple sources. When reviews are positive and specific, they can also be useful for motivating staff. The challenge for social media, communities, and review sites is the same as those for customer complaints and contact data: You must decide how to extrapolate the data to the marketplace as a whole.

Employee Input

There are two major types of employee input on the CE: employee surveys and information from employee input processes using email or instant messaging systems. Employees often see customer issues when customers see them. The question is how to inject this insight into the VOC process. Employee surveys are traditionally focused on employee happiness with pay, benefits, and supervision, with some attention paid to training and advancement. They are not very helpful in understanding the CE. However, a survey that asks employees about the specific issues that frustrate them—their causes, how often they arise, and how much time is wasted each time they occur—does provide valuable information.

Rapid issue input processes can be a great source for identifying emerging issues because they can often highlight problems not found in surveys and inspection mechanisms. For example, one airline (prior to a merger) had a feedback system that required the lead flight attendant to send an email within 30 minutes of landing that summarized the three top customer issues encountered on the flight. Three issues per flight times 2,000 flights a day provided 6,000 in-flight issue data points per day—almost in real time.

Summary of Strengths and Weaknesses

Each source of data describing the CE has its pros and cons. I recommend collecting and integrating all seven types of data in spite of the cons because each provides a different and unique perspective. Figure 7-2 summarizes their strengths and weaknesses.

Figure 7-2

Summary of Data Source Strengths and Weaknesses

Source	Strengths	Weaknesses
Internal Operational Metrics Transaction and system records of what the company did and did not do to/for the customer	Credible to management and useful in problem solving (to the degree that they describe factors that are important to the customer) because they are operations data	Provides a limited view of the customer experience based on only the aspects of operations that management measures (such as billing errors, late deliveries, etc.)
Internal Quality Metrics • Inspection data on defects • Call monitoring data • Service access data	• Allows identification of cause of original contact/problem • Provides data on effectiveness of service access and process	• Human review labor-intensive • Often focused on script compliance vs. broader issues • Speech analytics are expensive
Customer Contacts and Complaints Description of CE from customer perspective, including expectations and product use	• Very timely and descriptive of the actual customer experience • Provides root cause and emotional impact • Good source of positive feedback on employees	Data is fragmentary, unrepresentative, and must be extrapolated from the customer base
Mobile Transaction Data While growing rapidly, basically the same as contact, survey, and complaint data via other channels	• Like contact data, very timely • Volume increasing rapidly	Due to restricted input, often cryptic and incomplete
Customer Surveys Broad information on CE based on specific questions for relationship and specific transactions	• Data can be projected to the customer base and markets (with proper sampling), and ongoing comparable measurements are possible • Best analysis of drivers of loyalty	Significantly more costly and often less timely than data from internal metrics and customer contacts
Social Media, Reviews, and Communities • Public posting from small segment of total customer base • Community input can include thoughtful input from super-users	• Very timely feeback • Community members provide thoughtful input and reaction to company proposals	• Information incomplete and hard to get additional details from customers • Quality of data is variable
Employee Input Can be real-time via email as well as advisory boards and larger surveys	• Can identify process and customer-based causes • Can quantify amount of wasted effort due to problems	Surveys often not aimed at service; employees not given results of input to feedback mechanisms

To hear the voice of the customer, you have to question and listen to the employees involved in the processes that affect the CE. Comprehensive analyses of CE also show that the single best predictor of loyalty is whether the customer had a problem and how it was handled. Thus, data from customer service interactions describing problems must find its way into any VOC program worthy of the name.

In sum, the factors that create or erode customer satisfaction and loyalty are complex and therefore cannot be captured by any single method. Furthermore, every data source has its strengths and weaknesses. Thus, every organization that aims to build and maintain customer satisfaction and loyalty needs an effective VOC process that draws upon multiple sources of information.

Building VOC: Avoiding Four Common Implementation Challenges

Implementing a VOC process based on integrated data generated by multiple sources entails four challenges:

1. Gathering and classifying data so that it fits together
2. Creating a unified picture
3. Quantifying the implications
4. Compelling others across the organization to read and act on the reports.

Gathering the Data in One Place and Classifying Data

The VOC manager must take the lead in educating the IT department and the CRM vendor as to the requirements of a unified VOC. Both the IT department and the CRM vendor must understand that data should be collected from every transaction and customer touch noted in the CE process map discussed in Chapter 4. This allows the assembly of the fragmented sources of operations, quality improvement, call center reporting, and marketing services departments. The solution in most cases is to transmit all of the operational and mobile transaction data to the CRM system. Managing technology is addressed further in Chapter 8.

To the degree possible, coding must describe the symptom, problem, expectation, cause, and impact for each experience. Furthermore, information about the product involved, the type and identity of the customer, and the geographic location should be included. When there has been extended customer contact or investigation, it may also be possible to code information on root causes, actions taken, and the resulting impact on loyalty. Although the following classification scheme must be customized for a specific organization, it suggests the basic types of categories that an actionable approach will require:

- *Reason for contact/symptom*—From the customer's point of view, but in terms that are useful to the company
- *General cause*—In terms of customer error or expectation, a product defect, policy, or process or employee error
- *Root cause*—Often the specific cause remains unknown because investigating the cause of minor issues is inefficient (focus only on big issues)
- *Escalation code*—Why the issue was not resolved by the front line
- *Product and location identifier*—Possibly including production code
- *Action taken to resolve*—Often called disposition, which is not the same as root cause (The same disposition can apply to multiple causes.)
- *Outcome*—Satisfaction and loyalty based on follow-up surveys, call monitoring, or speech analytics

Each data element should be numerically coded to allow data processing and contribution to the analysis. For example, combining the symptom, cause, and product identifier can guide CSRs toward the appropriate response—and the analysis process can guide them toward the appropriate corrective action.

The codes used for describing the CE must be actionable, that is:

- *Comprehensive*—Covers the full range of issues across the CE

- *Granular*—Must be specific ("tastes too salty") rather than general ("tastes bad")

- *Mutually exclusive*—Overlapping categories will generate useless data (A utility company used "poor service" and "missed appointment" as separate categories, when the latter is a subset of the former.)

- *Hierarchical*—Codes selected from short drop-down menus with no more than 10 subcategories. (Good CRM and speech analytics systems can help with coding, but you must test them to ensure they meet these criteria.)

Creating a Unified Picture

The second major challenge is extrapolating the contact data and social media contacts to the entire customer base and then reconciling the data to the other operational and survey data sources (which are at least somewhat representative of the whole customer base). You must know what percentage of customers who experience a problem go to any particular service touch point or social media tool. The contact data multiplier, as described in Chapter 2 and earlier in this chapter, must be used to integrate data from touch points with survey, operations, and inspection data.

The best approach to quantifying the multiplier for key problems is to execute a one-time survey of a random sample of your entire customer base. Use this to determine the ratio of problems customers encounter to those reported to each company touch point or social media tool. Ask customers who have had problems which were most serious and whether they complained. If they did complain, ask where they complained (i.e., at which touch point). A sample result of a survey to understand where customers complain is provided in Figure 7-3.

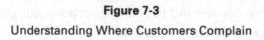

Figure 7-3

Understanding Where Customers Complain

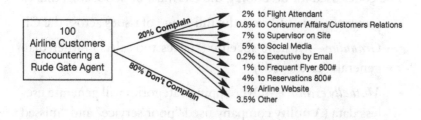

Quantifying the Revenue Implications of the CE

Once you have the unified estimate of the number of customers who have encountered a problem each month, you can then estimate the revenue impact with the Market Damage Model discussed in Chapter 3.

In the airline example, if an encounter with a rude employee damages loyalty by 25 percent and a customer is worth $2,000 per year, then you can conclude that 555 customers per month encountering rudeness will cost the airline $277,500 per month (555 × 0.25 × $2,000) or $3.3 million per year at just this one airport. In this way, you have taken six Web complaints, 20 social media mentions, and 14 calls to reservations and translated them into a multimillion-dollar problem that gets management's attention. When the revenue impact is made explicit, you have information that will motivate management to act. You can have even greater impact by also including WOM and risk impacts.

**How to Calculate Multipliers and Develop
Estimates of Problems in the Market**

Figure 7-3 illustrates that of 100 consumers who encountered a rude airline gate agent, only 20 complained: five customers went to a social media site, four called reservations (multiplier of 25:1), one went to the frequent flyer 800 number, and 1 out of every 500 sent an email to an executive. This 500:1 ratio is what we call the multiplier for executive complaints. Likewise, for complaints to reservations, the multiplier is 25:1; that is, there are 25 instances in the market for each complaint received at reservations.

GETTING STARTED
Questions to Ask Yourself About Your VOC Process

1. What are the most actionable survey and operational data sets that could be combined with customer contact data to address a recognized CE problem?

2. How can you connect these data sets to estimate the total customers in the marketplace experiencing that problem?

3. Can you use existing survey data to conservatively estimate the damage to loyalty when customers encounter a particular problem and thereby the revenue damage per month of all those problem occurrences?

4. Which of the three major strategies for addressing the issue you have just analyzed is best—prevention, surfacing more of the unarticulated problems, or better handling the problems already being reported by customers?

This exercise will give you a more powerful case for taking action on a recognized CE problem. Further, the answers to the questions will also show how fragmented your data and classification schemes are and uncover opportunities for quick fixes. The answers to these questions will give you a foundation on which to build a truly effective and efficient VOC process—one that more than pays for itself!

KEY TAKEAWAYS

- The VOC should identify emerging issues, significant new issues that management is not aware of, and progress on issues that management believes are currently being addressed.
- The VOC must include data from surveys, contacts, operational transactions and failures, and employee and social media input to allow estimation of the volumes of each issue, as well as the impact of the experience on loyalty and WOM.

- The data must be integrated to create a unified picture. The key is to extrapolate data from each touch point to arrive at a best estimate of the number of customers who had a problem in the last month.

- Effective use of both CRM and voice analytic tools is critical for managing the avalanche of mobile, social media, email, and website inputs. The VOC manager must actively manage both in-house and vendor technology to ensure that the same customer identifier is used everywhere and to produce an integrated, end-to-end picture of the CE.

- There must be one person assigned to coordinate or at least facilitate the VOC collection, analysis, and selection of issue priorities based on economic impact. Otherwise, VOC will have little impact. The most cost-effective collection is usually decentralized, but the most effective analysis is centralized and executed by senior analysts with operations experience who then keep track of which issues are fixed.

- The VOC must identify only a few issues for any manager to address, ensure that manager's understanding of the issue and possible solutions, and create an economic imperative for action.

Notes

1. John Goodman, Cindy Grimm, and Joshua Hearne, "Improving the Customer Experience," *Call Center Pipeline* (Annapolis, MD: Pipeline Publishing, 2012), pp. 28–30.

2. Conversation with Matt Trifiro, then senior vice president of marketing at Salesforce.com, November 2012.

3. Intelliresponse, "Consumer Use of Social Media for Customer Service," White Paper, Toronto, 2011.

4. Scott Broetzmann and Marc Grainer, *2011 National Rage Study* (Alexandria, VA: Customer Care Measurement & Consulting, 2011).

5. Scott Broetzmann, Marc Grainer, and John Goodman, *2013 National Rage Study* (Alexandria, VA: Customer Care Measurement & Consulting, 2013).

6. Eduardo Laveglia, *Complaint Behavior in Argentina* (Buenos Aires: Proaxion, 2012).

7. Interview with Melody Blumenschein, Digital Media Manager, General Motors, December 12, 2013.

3

Key Issues of Implementation

CHAPTER 8

Taming Technology

I recently called my Internet provider because the broadband was down. After plowing through six levels of menus, entering my phone number twice, and indicating what kind of service I had (why did they not know from the phone number?), I was finally asked what my problem was. When I responded that the problem was connecting to the Internet, an on-hold message then informed me that most problems could be resolved more quickly on the company's website. The Internet provider's use of technology seemed designed to turn the customer into a raving maniac. Do they do this on purpose?

Similarly, your company may have invested millions of dollars in service technologies. Results, however, come down to how those technologies are configured. Today's fast-emerging mobile and Web capabilities, in particular, can empower customers to service themselves, whenever and wherever they want. When technology is poorly implemented or used, it becomes a major barrier and a part of the problem rather than a path to resolution. Taming technology means making the technology transparent to the customer and using it to seamlessly deliver the product or valued service that the customer desires. In this chapter, you will learn:

- How to align technology and the ideal CE.

- How to smooth the impact of technological evolution.

- The benefits, pitfalls, and best practices of the most prevalent technologies.
- Metrics that will allow you to monitor technological effectiveness.

Full disclosure: This is not a chapter on IT project management. Books and entire management degrees are available on that topic. This chapter aims to provide tools and the understanding to harmonize the evolution of technology and the CE.

Aligning Technology with the Ideal CE

A process map of the current CE should be your primary guide for applying technology. Once the cross-functional CE process improvement team has mapped the current CE, many process disruptors and unnecessary activities that create cost and dissatisfaction will become obvious targets for improvement. This mapping helps avoid the two biggest challenges in implementing technology: automating existing non-ideal, wasteful processes and modifying the processes to fit the technology.

Constructing the CE Process Map for Existing Processes

The CE map of the current process can be constructed using the key activities contained in the four-part CE framework in Chapter 2: DIRFT, access, service, and listen-and-learn activities. Figure 8-1 lists the 12 activities for the end-to-end CE, using the eight more detailed activities contained in DIRFT, as described in Chapter 4.

Once the 12 boxes are arrayed, your next step, in conjunction with the frontline staff and managers, is to map the detailed process for each of the activities. For instance, the purchase activity should describe the flow of each way a customer can purchase the product, including what happens if they change their mind, make mistakes, or use a credit card that is rejected. This process of mapping current CE

Figure 8-1

Major Activities to Be Included in Process Map of Ideal CE

Key CE Activities to Be Process-Mapped	
Product Design	Start-up
Marketing	Ongoing Use and Billing
Information Search	Providing Access to Service
Sales Offer	Customer Service
Purchase	Listen and Learn
Delivery	Feedback on Listen and Learn

is identical to mapping the customer journey as described in the *Harvard Business Review* article, "The Truth About Customer Experience," by Rawson, Duncan, and Jones.[1] The result of this mapping exercise will most likely be a flowchart many feet long, with dozens of activities for each of the 12 basic activities (I have seen complete CE charts 40 feet in length—you just work with one activity area at a time). Each activity will interact with previous and future activities. For instance, Figure 8-2 shows the primary process flow within the service part of the CE framework; each activity could and should be further detailed to create the more detailed picture of the customer journey across all the touch points.

You will note in Figure 8-2 that the activities of creating awareness (activity 1) and proactive communication (activity 16) are actually part of the service access and DIRFT parts of the CE framework, respectively. However, both of these communications are often initiated by activities in the service part of the framework. Similarly, activities 13 and 14 (statistical generation and policy analysis for prevention) are components of the listen-and-learn part of the CE framework. Their performance is often embedded in the customer service function. Analysis of these interlocking flows will highlight delays and sources of error.

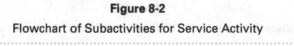

Figure 8-2

Flowchart of Subactivities for Service Activity

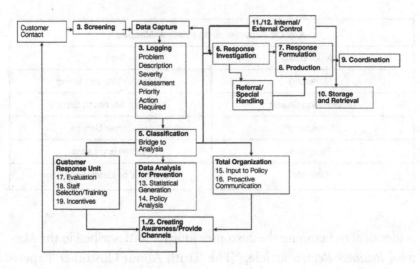

Review Existing Process for Disruptors

Once your team has mapped the current CE process, the next step is to identify opportunities for process improvement. The best approach is to examine each part of the process to identify the causes of customer and employee frustration, unmet expectations, errors, delays, and extra effort. The CE process improvement team can look for several reliable symptoms of broken processes. These symptoms include rework (e.g., where business forms or transactions are returned to be done over or the same employee or customer must redo the activity due to an error), ponderous manual actions by the customer or employee, dropped calls, or a delay while awaiting information or decisions. Whenever the company returns a submission to the customer or a customer returns or rejects a product or offer, there is probably an opportunity to improve the process.

For example, when an insurance underwriter found an error in an application, she stopped her review and returned the form to the customer or agent for correction. That person fixed the error and re-submitted the application, which then required a whole new entry in the application logging and tracking system. It then went back to the original underwriter, who reviewed it again. If she found a second error, she rejected the form again, and the same process was followed. This sometimes happened four times on the same application, causing delay, frustration, and expense.

Whenever the process chart shows delays, failed transactions, or returned work, extra cost and unnecessary dissatisfaction result. A good process fix is to create a form online that educates the customer on common mistakes and then executes an edit check as the form is completed so that an erroneous form can never be submitted.

Map the Ideal CE Process

Once the process disruptors have been identified, at least at a macro level, go back and map what the ideal process should be. For this activity, the best consultants are your frontline staff who actually do the work, as well as a focus group of customers. You often can recruit willing customers to give you input from your customer complaint database. The ideal CE map should be your compass for the future.

Select Process Glitches to be Reengineered

The opportunities for improving the current CE should be triaged to set priorities and focus on a few of the most flawed aspects of the current process. Survey and complaint data will be helpful in setting such priorities by identifying the customers' key points of pain. Especially at the beginning, you should pick and attack one or two issues. Better a small success than a big disaster.

Assess any Proposed Technology to Assess Impact on CE

Once maps of the current and ideal CE process are in place, conduct a review of any proposed new tool or technology to identify how it will change each phase of the current CE process and whether it will move the company toward or away from the ideal process.

Communicate to the Organization and Customer Base

A major error in many organizations is waiting until the process has been reengineered and the programming completed before telling the rest of the organization and the customer base what is going on. This lack of communication both precludes useful input and leaves everyone with the impression that management thinks the current CE processes are fine, when both employees and customers are actually frustrated. A bank used a proactive approach when it broadcast upcoming systems changes and transitions well in advance to customers, as well as to employees. This approach significantly improved the perception of both caring and competence. The marketing department often discourages such communication on the grounds that it implies that the current process has problems. In reality, everyone knows there are problems. Customers and employees need and want to know that the problems are being addressed.

Smoothing the Impact of Technological Evolution

The ideal CE map should be the basis of your discussion with the IT department. As soon as the IT, marketing, CE, and operating lines of business units are all working from the same agreed-on reality, your chances of success improve. Even though the ideal CE map provides the basis for communication, you now must actively facilitate the communication. The following are some best practices for making the evolution of technology a positive experience for all parties, internal and external.

Require That the Same Customer Identifier Be Used Everywhere

Not all data and information must be in the same system; it just has to be linkable and based on uniform definitions. This requirement has four aspects. First and most important, the key to linkage is the customer identifier: The same customer identifier must be used in all databases. Second, all accounts belonging to the same customer and/or household must be householded, that is, associated with each other. Third, a customer identifier should be tied to all operational transactions. Finally, operational transactions and customer issues must all be defined the same way. Engineers cannot refer to the issue as corrosion while the service area refers to it as surface discoloration. If these four aspects of data management are in place, there is a high probability of what is now called Big Data success. The Cheesecake Factory has executed this brilliantly. It used a single meal order and billing system to tie together customer complaints, meal process inspections, kitchen employment, and improvement tracking to each of the 80 million meals served annually.

Jim Albert, CIO of Bankers Financial Corporation in St. Petersburg, Florida, has developed several strategies to mitigate the different perspectives of the IT and marketing departments and executives in charge of lines of business, CE, and operations.[2] Several of these strategies are now discussed.

Blur the Lines

At Bankers Financial, IT business analysts work with the marketing department to develop both business requirements and IT design specifications. Further, the departments must make joint presentations to executive management and often have joint happy hours. Finally, IT departmental staff and executives are required to periodically observe how the systems are used by the agents (who are the external partners), the internal operations employees, and customer service frontline employees. The IT staff and executives, while observing, often ask the employees, "Why did you do that?" Such anthropological observation works better than focus groups and standard interviews because the IT staff and executives see what is actually done versus being told what the staff knows is supposed to be done.

Solution Design Thinking

Move away from the megaproject and focus on day-to-day operations to address practical problems. Avoid the we-need-a-completely-new-system thinking and ask what can be done now to improve the current system. This approach resonates with me because I recently heard an IT executive tell a line customer service manager that the module she desperately needed would be delivered "in about three years." Jim Albert has instituted what he calls, "Do it! Week" where a total fix to an operational frustration must be analyzed, developed, tested, and implemented within a single week. He says it sometimes takes less than one day from start to finish to make a small system change that can have a significant impact on the CE.

Pilot-Test Process Changes

The best people to pilot-test an activity are the people who will use it, either employees or customers. IT and CE should partner to pilot the improved processes manually to identify glitches. Then the new technology should be tested in a real but safe environment. This is how companies such as Chick-fil-A, CVS Pharmacy, and Family Express (the Illinois convenience store chain) test new concepts—in a store that is like a laboratory but using real employees and customers. A quasi-laboratory setting allows the measurement of all aspects of the experiment and openness to on-the-fly changes. The application development can be streamlined using tools available in The Cloud. Also, the risk of a disruption due to a development error can be limited by testing the change in a single small product area or geographic territory.

After-Action Analysis

Every IT implementation should be analyzed to identify problems so that similar future implementations eliminate the problem. A major bank found that the same so-called standard problems always occurred during the system transitions when a new bank was acquired. Once the source of these problems was identified, part of the system transition plan for each new bank acquisition was to plan so that the standard problems were eliminated and did not happen. The bank went from losing a significant number of customers during system transitions to actually gaining customers during transitions.

Celebrate the Improvements

Recognition is a key motivator, and so two actions are required here. First, celebrate the success of the whole team including the IT, business improvement, and operations staff. When employees receive positive executive feedback in person, it makes a huge impression, and they will try to repeat their behavior. Second, communicate the successes to all employees as well as to customers. I briefed an executive on some ongoing problems that both customers and employees had reported, and he retorted, "But we fixed those problems last quarter, and the fixes have been cascading down the organization for the last two months." I asked, "Have you broadcast and celebrated the fixes?" They had not: The company had assumed that employees and customers would see the changes. It can take many months for customers, and even some employees, to perceive a change unless you point it out.

Benefits, Pitfalls, and Best Practices of Available Technologies

All of the following technologies will benefit the delivery of a great CE with high efficiency. However, each technology also has serious pitfalls that turn it from a source of efficiency and satisfaction into a maddening point of pain. Therefore, the CE leader must encourage the aggressive use of all the following technologies in order to achieve all the benefits, while avoiding the inherent pitfalls.

CRM Systems

CRM systems must be the CE management hub. To do this the CRM systems must be equipped to facilitate all four of the following actions:

1. Manage and guide the response content and process of all customer interactions. Contact management systems primarily focus only on this objective.

2. Anticipate and take action on customer needs by drawing on internal information about each stage of the customer's lifecycle, as well as stated preferences, and on all operations information and external data sources, such as social media aggregators.

3. Obtain (ideally automatically) information from all customer touch points (everything from account numbers and orders to survey feedback) to support response, emotional connection, education, and input into VOC.

4. Obtain information from the Knowledge Management System (KMS) and operations data for use in response and proactive communication.

You must use the same customer identifier across the company in all systems. If you do not, you have no hope of effective CE management. Not using the same customer identifier in all systems is the most common mistake in implementing CRM systems. The second most prevalent mistake is selecting an off-the-shelf CRM system and not modifying it to meet your ideal CE or, worse, changing the CE to fit the CRM system's existing, inflexible processes.

Do not rely on vendors' promises. Vendors will often say, "Our system can do that!" Demand to see where it has already been done in exactly the desired way and gain an understanding of the effort required to "make it do that."

Two more tactical errors made when implementing a CRM system are:

1. *An inability to connect CE to existing operations failure data*—For the data to be useful, it has to be collected and reported in a disaggregate format. For example, you need to know more information other than 1,000 incorrect service charges were levied. You must know which customers had that problem.

2. *A weak reporting capability*—The system cannot preclude the easy analysis of the content of customer transactions and process parameters (such as telephone call talk time or email response times), and it should allow unified analysis and reporting of quality, content, and operational parameters.

A best practice for CRM is to use transaction failure data from operational databases to proactively warn customers in advance about impending problems or unpleasant surprises. For example, a men's clothing chain on the East Coast uses customers' zip code data to identify those on the West Coast and email them that delivery will take two extra days. Even though customers seldom read the delivery standards on the website or in the order, they do read short emails about delivery times.

Email and Chat

Email and chat are now becoming integral parts of CRM systems. They are often driven by speech or text recognition systems and supported by automated intelligence (AI) tools to suggest the best responses to service reps or provide them directly to customers. The technology is now able to select key words and, by reviewing multiple phrases together, to recognize intentionality, that is, why the customer is asking a particular question or using a phrase. For example, "Where is this product made?" could imply a concern about delivery delay, a product quality, or product content issue.

Misunderstanding a customer's email or chat request leads to inappropriate responses or ineffective triage. Even when the customer's inbound query is not detailed enough, or when the text recognition tool is not exact enough for the query to be clear, most automated response processes keep trying to respond even after the customer has expressed frustration with the answer. A smart tool, powered by AI, will quickly acknowledge that it has failed and transition the customer to a human. Another approach is for a recognition tool to recognize ambiguity and ask two more diagnostic questions in order to avoid providing an array of inaccurate responses.

In email, a best practice is to ask customers how soon they need a response. Most companies avoid asking this question because they are afraid that the answer will be, "Right now." However, it is better to know and maybe reset expectations than not to know. Also, most customers are reasonable and will be happy with a several-hour delay.

A best practice for chat is to push an automated offer of chat support in response to inaction on the customer's part, such as sitting on a page for 90 seconds. When customers then say they want to chat, you have no more than 45–60 seconds to connect them to a live CSR.

Website

The website must be treated as the first and primary communication channel with the customer. Almost everyone goes to the website before buying or obtaining service. Almost everyone with a question or problem goes to the website before emailing or calling.

The biggest errors made in website design are:

- Treating the site primarily as a marketing channel when almost all websites are most often visited by existing customers looking for help.

- Making it hard for visitors to navigate due to a poor website map and lack of an orientation section for new visitors.

- Making customers log in with a password and user name when all they want is general information. This creates an additional barrier. Provide a robust, well-indexed FAQ section outside the firewall.

- Providing a website search engine that does not immediately take a customer to a page with the key help links and answers to top issues, as well as a multichannel Contact Us mechanism when the words *problem* or *complaint* are entered into the search box. In many websites, such a search leads customers to directions for submitting a regulatory complaint—the worst possible outcome.

Best practices for websites are:

- The website map is an index to all the issues customers will want to ask about and is no more than four columns of customer-friendly terminology. Its content *does not* go below the fold of the screen. This map is highlighted at the top of every page, not in the fine print at the bottom. The website map on a mobile application should be driven by a search box where customers can enter terms or issues and be taken directly to the information.

- Include the responses to the five most prevalent questions/issues being received in the contact center on the website home page, not five clicks into the website. Ideally, this list is a living list of issues that is updated weekly—that is, this week's list is based on the calls received last week.

- Include all communication channels prominently on every page. If customers have an unpleasant surprise from the current channel, make it easy for them to shift to another preferred channel. For example, I am happy to use the self-service tracking system until I find that the package has *not* been delivered. When that happens, I want to talk to someone and need a Call Me button or an 800 number. Providing only the channel you want the customer to use results in abandonment, dissatisfaction, market damage, and lower contact rates.

Speech/Text Recognition and Analytics

Speech/text recognition and analysis tools allow the conversion of unstructured information into structured knowledge. However, there are three primary pitfalls in their implementation. First, some tools primarily depend only on key word searches and do not reliably and effectively recognize key CE events and customer intentions. Advanced systems analyze phrases and the location of subjects and verbs to identify meaning and intentionality. Second, many lack effective identification and reporting of recognition errors. Third, most user companies lack qualified staff to continuously tune the system.

A best practice for recognition systems is to obtain customer feedback on every interaction, using the recognition system to allow identification of failures. This allows the company to rigorously analyze the causes of error and dissatisfaction, and continuously tune the system. Another best practice is for the recognition system, after it has failed twice in a transaction, to pass the customer to a human.

Interactive Voice Response/Automated Call Directors (IVR/ACD)

The IVR/ACD configuration must perform the intake function described in Chapter 6 in a transparent manner that minimizes confusion and misdirection. Menu selections must be simple and be part of the call-in information provided in the literature or on the website. Additionally, modern ACDs will identify the customer's phone number via automated number identification and, if the queue is more than one minute, offer the customer a virtual queue; that is, the company calls customers back when their phone numbers reach the top of the call queue.

Five major mistakes are made with IVR/ACDs:

1. The IVR/ACDs are too complex and do not provide maps of the process, so customers get lost in it.

2. More than four options are provided without providing a map of the menu and/or more than three levels of menus are used.

3. The message begins "for faster service, go to website" and "our options have changed." Both are irritating because most customers have gone to the website before calling, and almost all IVRs say the options have changed when they have not.

4. The ACD neither captures the phone number nor uses computer telephone integration with the CRM to identify the customer proactively to the CSR.

5. The ACD does not feed talk time and CSR wrap-up time into the CRM system.

Four best practices for IVR/ACDs are to:

1. Print the menu wherever the 800 number is provided.

2. Allow the customer to enter information such as account numbers while in queue.

3. Link the ACD and any customer-entered information to the CRM system.

4. Use a virtual queue tool whenever the queue goes longer than two minutes.

Video

Video is one of the most powerful technologies for educating customers and preventing problems, and it helps companies connect with customers. A picture truly is worth a thousand words when it comes to CE.

For instance, a drug company shows short videos of customers who have quit smoking, describing how they went about quitting and the challenges they encountered. The videos show a range of diverse speakers so that viewers can identify with at least one of them and accept their advice. These videos convey information and allow emotional connection and identification with the customer. Further, a 45-second video tutorial can teach a customer (especially one under 30 years old) to use a function more effectively than a paragraph of text. Video servicing can allow connection, but not everyone wants to be seen by the CSR.

The biggest error is to use videos that are too long and/or too slick. The most effective videos are usually less than a minute long and are often made by your frontline employees who are much more genuine and credible than actors. For example, Lowe's home improvement store uses six-second videos (titled *Fix in Six*) to effectively communicate useful home repair tips.

Knowledge Management Systems

KMSs are critical to effective cross-channel response and the proactive education of customers. A good KMS creates the capability for, as Thomas Davenport calls it, "knowledge capture and sharing for a more effective collective intelligence."[3] This collective intelligence can dramatically reduce training time and actually reduce customer service workload if the database (e.g., common troubleshooting methods and error code translations) is made available to customers via the website and the database is partitioned (e.g., divided into employee-only view and customer/employee view).

The biggest problem with KMS is that companies don't maintain the information in the system and fail to share it with customers. The KMS should be a resource for both internal and external customers; it is a core repository of knowledge. Even in small organizations, at least one full-time person should be maintaining the KMS. This investment will be more than offset by savings in service cost because customers will be able to find their own answers if the KMS is partitioned and the bulk of the database is made available on the website.

A best practice for KMS is to continuously update the information in the database via input from the frontline staff based on their research, resolution of new problems, and knowledge of issues. At Harley Davidson, the service manager encourages frontline staff to pursue new issues to resolution and to report the issue in a standard format to the KMS administrators for final formatting and entry into the KMS.[4] He also recognizes and congratulates the team members who make the most input. An additional best practice is to monitor the effectiveness of the KMS by team members to identify both employees who need more training and response topics that appear ineffective.

Mobile Technology

Smartphones and mobile technology are now the most used link to the company's website, phone, email, and video systems. Mobile technology also includes transaction processing capability, as well as access to Web-based review sites and their online communities. If the company's capabilities disappoint, the review sites and communities are immediately available for complaint.

Two trends have huge implications for mobile technology. First, mobile technology has made the Internet available to consumers who cannot afford desktop computers or laptops and who do not even have electricity or a permanent place to live. Villages in rural Egypt and Chad use mobile phones to pay bills, order goods, and interact in social media. For this to work, all website functionality must be simplified to work on the phone.

The second trend is for IVR and speech recognition technology to supplant text communications and even the use of keyboards. Soon, typing on keyboards will be viewed as way too much effort and a major source of errors. And, more importantly, illiterate consumers will be able to use a wider range of Web functionality even though they cannot read. For instance, Millicom, which operates the Tigo cellular system, provides an IVR-driven financial services system (e.g., bill paying, money transfers) in five languages that can be completely operated using only numbers 0–9.

The biggest pitfalls with mobile technology are the company's failure to translate functionality to the small screen and the failure to make channel switching easy, such as from website to phone. This is not so much in order to make reading easy as to make fat-fingering less likely. Now that buttons have evolved to touch screens, the functions must be large enough to avoid errors that frustrate even more. The failure to offer calling capability usually stems from the fear that if calling is offered, the customer will actually use it, thereby driving up costs. As noted multiple times: Better to hear from the customer than not hear.

Social Media (Other Than Online Communities)

Social media are usually a place for friends and colleagues to exchange social information. You could argue that each person's Facebook page is their community or the front yard of their home. In any case, individual social media is like someone's home and very different from formal online communities. Companies should tread lightly in someone else's front yard and enter only when invited or given permission.

That being said, if customers are yelling good or bad things about you in their front yards, it is OK for you to listen via monitoring tools. The next section discusses communities.

Online Communities and Review Sites

Online communities and review websites are both important tools for companies, even though businesses cannot totally control these sites. User community websites can provide mutual support for customers and are a great source of both VOC and customer engagement, but they are beyond the company's control. Some companies are creating their own review websites, but even here, the company cannot completely control them.

Companies must make four interrelated decisions about online communities that affect the extent of community impact on CE. These decisions address its objectives, how it will be hosted, its functions, and how to motivate member participation.

The first decision is whether the primary objective of the online community is to retain existing customers or to win new ones. The second decision is whether the community will be positioned on a private company-hosted website or on a public social media platform, such as Facebook. If the primary objective is to retain existing customers by being a category expert and providing customer support, a privately hosted community is better. If the objective is to attract new customers by creating buzz and large amounts of word of mouse among consumers, then use a public platform. For example, on Facebook, members can easily and automatically pass on items to their friends. The more specialized your target population is (e.g., accountants or engineers), the less useful a general public platform is, even though hosting is technologically easier on a public platform. It is also possible to start on a private platform and create links to and from public platforms.

Secondary objectives, such as idea generation and doing social good, can be achieved through either platform because each contributes to both primary objectives. Having ideas accepted and creating feel-good by doing good deeds make current customers happy and generate a lot of buzz.

The third decision is the range of functions you want the online community to perform. A blog with readers who comment is a very simple online community at one end of the spectrum of functional complexity. Figure 8-3 provides a spectrum of possible community functionality, going from the very simple to the very complex.

Figure 8-3

Possible Functionality for an Online Community, from Simple to Complex

An important implication of this spectrum is that a company can start an online community with a simple blog and scale it up gradually. If you can find the right blogger with good content and an engaging, entertaining personality, you will be able to quickly add additional functions such as support and member-contributed content. An enhanced community will then create buzz among your specialized customer base and attract new members and potential customers.

The final challenge is probably the most difficult: motivating members to participate. Three factors lead to participation: content, fun, and incentives. For company-hosted online communities aimed at retaining existing customers, content is probably king. Consumers want information and advice in their interest area, whether it is hot cars or child care. Business customers want support and timely expert information that will help them succeed. However, both groups also want fun.

Finally, fun content is at least entertaining and ideally makes you laugh. The online community moderator must have a naturally enthusiastic personality and really enjoy moderating. In a good example, Goodwill of Greater Washington, D.C., which sells used clothing to support job training programs, created the blogger and moderator position of D.C. Goodwill Fashionista. The online community, which started as a blog, quickly attracted over 5,000 followers—contributing to a significant increase in store sales. The D.C. Goodwill Fashionista Blogger, Lisa Rowen, finds that self-depreciation is engaging and disarming. She has published pictures of herself in outfits she decided were a mistake with the heading, "What was I thinking?" Fun is critical for consumers; this is the reason Old Spice aftershave commercials have been viewed 250 million times on YouTube. Regardless of your audience, you must have a sense of humor and occasionally be silly.

The third motivation success factor is incentive. Incentives can be intangible recognition (having your picture on the corporate Facebook page or website) or can be tangible, such as a free sandwich. For example, members of Chick-fil-A's multimillion-member online community dress their children and babies up as cows and go to the restaurant on special dress-up days both to have their picture taken for the community website and to get free sandwiches. In another example, at the specialized technology website community, Stack Overflow, being voted by members as having provided the best technical solution is a strong incentive. A final example is Starbucks, which uses a technique called gamification (to be discussed), feeding back congratulatory messages to Starbuck online members who submit ideas that are accepted.

The biggest pitfall of company-sponsored online communities is when companies do not invest in the monitoring and support of the community. The worst thing you can do is throw a party and have lame refreshments and no music; people walk in and immediately walk out.

Three other big errors companies make are failing to analyze the comments and reviews in aggregate (for policy purposes versus individual situations), failing to quickly respond (within two hours) to legitimate criticism when appropriate, and attempting to suppress controversy. Analysis of the content of reviews is an ongoing task that text analytics does well. On the other hand, such tools are still somewhat expensive, and a daily visual scan can pick up most important issues. This same daily scan will identify issues that need action, either offering to fix the situation for the customer or responding in a low-key manner. A good online community management system will allow identification of negative comments and provide an interim response that buys time while appearing responsive.

A final pitfall is trying to suppress controversy. Respectful controversy stimulates interest and discussion, and moderators should allow it. Joy Montefusco, the Interactive Executive Producer of the Discovery Channel's online communities, allows each community to set the standards. The moderator intervenes only after three community members flag a comment as inappropriate.

A Word About Gamification

Online communities are also a perfect environment for the use of gamification as an incentive. Gamification capitalizes on people's natural competitive nature and desire for achievement and recognition. Someone who does something you like gets points, and if they do it enough, they get a virtual award or badge. Often, the points or badge have no value other than recognition from the rest of the community that someone has done that activity more often. For instance, FourSquare recognizes the person who checks in the most at a store (or even a dog park) as the mayor of that location. There is no tangible value except recognition.

This recognition can be applied to employees, business partners, support communities, and customers. One company recognizes business partners who reduced the percentage of applications with errors by giving red, yellow, or green public ratings. The green partners were recognized, and nothing was said about the rest, but everyone knew who was where on the list. Intuit recognizes accounting software users who give the most support to other users. The fact that even accountants who focus on tangible assets can be motivated by SuperStar labels shows that recognition works almost everywhere.

Companies such as Boyd Gaming are motivating gaming customers using tools such as Gigya to both recognize top customers and provide the customer with tangible benefits (room upgrades and meals) and intangible benefits (status indicators). In short, gamification can be an effective motivational tool for employees, partners, communities, and customers but requires intense management. Although many are skeptical of the impact of intangible benefits, it certainly seems to work.

Metrics for Managing the CE Aspects of Technology

The important metrics to make technology a facilitator rather than a barrier to CE are all couched from the customer's perspective. Some of these metrics can be gathered internally from operations data, and some require feedback from one of two customer sets: your external customers and your internal customers—your employees. The following are the five major metrics you should consider:

1. *Percentage of transactions where the customer self-services by type of transaction*—For any transaction that can be self-serviced, at least 75 percent of customers will want to use self-service for its convenience. If the percentage is less than that, either the process is clunky or people are unaware of its existence.

2. *Percentage of transactions that are anticipated and executed before the customer asks*—The CE department, working with the IT department, should identify the transactions that are appropriate for anticipatory action (that is, psychic pizza) and set a target for both identifying and responding to these opportu-

nities. The savings from the avoided service expense, as well as the impact on satisfaction, should also be measured.

3. *Percentage of self-service transactions that fail*—The number of self-service transactions attempted by customers should be captured, and the percentage that fail, by type of transaction, should also be captured and analyzed. For example, when a search returns an answer of "No Records Found," your KMS has failed and you should track those occasions. Likewise, a negative answer to the question, "Was this FAQ answer helpful?" signifies a failure. This should include IVR transactions. In all cases, the best diagnostic data will be produced if the system identifies exactly where in the transaction flow the process failed and whether the customer simply abandoned the transaction or the system ended the transaction.

4. *Customer satisfaction with each of the service transactions*—This metric must be measured using a customer satisfaction survey. The customer should be presented with an opportunity to give feedback ideally within the transaction, especially if the customer abandons it. The survey should be a simple five-point satisfaction scale, followed by an open-ended question asking for the reason for the specific rating. I recommend a five-point scale with a neutral midpoint because it will provide enough granularity in the data and is less intimidating than a ten-point scale.

5. *Employee satisfaction with the transactions by type of transaction*—Employees should be asked to provide feedback on frustrations with service transactions that they execute on behalf of customers or internal customers. The best way to gather feedback is a quarterly employee survey by type of transaction. The survey should ask about failure rates for major categories and the employees' nomination for the two most time-wasting/frustrating transactions they execute.

GETTING STARTED

Questions to Ask Yourself About How You Use Technology

1. Have you constructed a process map of the ideal CE?

2. Have you evaluated every major customer interaction to identify where technology can help you anticipate or prevent transactions?

3. Does your technology facilitate emotional connection by providing the service employee with information about the customer history and circumstances?

4. Do you require that the same customer identifier and issue descriptors be used in all databases across the company?

5. Do you have metrics that show how often technology fails in servicing the customer?

KEY TAKEAWAYS

- The ideal CE process map, facilitated by process improvement, is the mechanism that unifies the efforts of the CE, marketing, and IT departments.

- Technology implementation should always require the use of a universal customer identifier and should always be pilot-tested with provision for obtaining detailed customer and employee feedback.

- The benefits of technology are anticipation, flexible channels, reduced workload, reduced customer and employee effort, the ability to emotionally connect, and effective VOC.

- Major errors in using technology in the CE environment include not providing employees and customers with the same information via the same systems, not recognizing that the website is the first service channel used by most customers, and not starting small for online communities or pilot testing other technologies prior to rollout.

- The key metrics for technology are the percentage of failed transactions and the percentage of service transactions where customers attempt to self-service.

Notes

1. Alex Rawson, Ewan Duncan, and Conor Jones, "The Truth About Customer Experience," *Harvard Business Review*, September 2013, p. 90.

2. Interview with Jim Albert, CIO, Bankers Financial Inc., November 18, 2013.

3. Thomas H. Davenport, "Enterprise 2.0, The New Knowledge Management," *HBS Blog Network*, February 19, 2008.

4. John Goodman and Crystal Collier, "Skills-Based Routing Versus Universal Rep, Which Is Best?" *Call Center Pipeline*, January 2013.

CHAPTER 9

Building a Culture of Empowerment and Connection

I recently arrived at the airport a bit early to check in for a flight and was chagrined to see that it was running over two hours late. There was no earlier flight, and I expressed concern about being late for a meeting. The counter agent looked at his computer and, without missing a beat, said, "You're a good customer. Let me see if I can get you on another flight that's running on time."

He then handed me a boarding pass for a competitor's flight. This did not immediately help his airline's revenue stream, but it certainly reinforced my loyalty to it. The agent's short-term investment bought a lot of goodwill and positive WOM.

This chapter addresses how to create a culture in which employees have the freedom and time to break the rules, use their empowerment, and leverage technology to create memorable, cost-effective emotional connections.

You will learn how to:

- Establish an environment that fosters empowerment and connection and that brings out the best in people.

- Plan for and create connection.

- Structure the behavior and role of executives and supervisors to set the proper tone and empower staff to satisfy and connect with customers.

- Measure and manage empowerment and connection.

In each of these areas, you will learn best practices and common pitfalls.

Establishing an Environment That Fosters Empowerment and Connection

An environment that fosters empowerment and connection must be carefully planned. The planning includes getting the right people on staff and providing the tools and opportunity for the employees to use their empowerment to satisfy and connect with the customer. Empowerment is the precursor of emotional connection. Genuine emotional connection cannot be robotic and must be discretionary based on the full range of circumstances observed by the employee.

Connection does not just happen. It is a conscious action during which employees assess the situation and the person in front of them (or on the phone or chat board) and then decides whether connection is both possible and desirable. For this to happen, the company must do two things.

First, the firm must create opportunities for employees to make these connections. PetSmart, a retailer with 1,300 stores, realized that buying a bag of dog food presented fewer opportunities for possible, desirable connections than dog training or grooming sessions. They also found that margins on these transactions were twice what they were on basic retail sales.[1]

Second, the employee must have the tools and skills to take advantage of the opportunity. We begin by addressing the prerequisites for that empowerment: assembling the people and tools.

There are six prerequisites for empowerment: the right people; an effective job description; guidance and training on handling the most prevalent difficult situations; information on the customer's situation; time to create a connection; and positive reinforcement.

Right People

Every company with a great CE acknowledges that the business is successful because it recruits and retains the right people. At Pet-Smart, for example, pet ownership is emphasized during hiring. The theory is that an employee with a pet can better identify and connect with customers. In another example, the Pointe Hilton Tapatio Cliffs Resort (Pointe Hilton), in Phoenix, Arizona, has impressed me with consistently superior service that parallels or exceeds what I have encountered at higher-priced luxury hotels. The hotel hires employees who demonstrate customer focus and who can explain how they have successfully handled challenging service situations in the past. Even the housekeeping and groundskeeping staff are treated as frontline service staff.

Job Description

The job description of frontline employees is usually very operational and touches upon CE only by requiring courtesy and responsiveness. An effective job description will explicitly create accountability for the use of empowerment, the creation of an emotional connection, and input into the VOC process. Conners and Smith describe creating accountability as "defining results that will get you the desired results."[2] If you want connection, you put it into the job description. The Pointe Hilton culture statement states that employees must anticipate the customer's needs and should ensure that the guest will return and recommend the resort. It states, "We trust your judgment—we will stand behind your decision" and "The only wrong thing you can do is to do nothing." At Chick-fil-A, they have greeters in the dining room at peak business hours whose job description is to "create microbursts of emotional connection" via 30- to 60-second interactions.

Guidance and Training

Guidance and training on how to handle specific difficult issues is critical. Most companies give general training on how to handle difficult customers but do not give the in-depth substantive guidance on handling the top five or ten specific types of issues. The guidance should include flexible policies (and the necessary training and role-playing on executing each of the policies) that result in the employee feeling confident enough to handle 99 percent of all the issues they encounter. The employee should first ask the customer, "What would you like me to do or what do you want?" so that the solution can be tailored to their needs.

The flexible policies should be couched in the form of flexible solution spaces that address the five to ten most prevalent, difficult issues that employees encounter with customers. A flexible solution space outlines multiple solutions (usually no more than four) to the same general issue. It is accompanied by the range of factors that the employee should consider when deciding how to select the appropriate solution and negotiate a resolution acceptable to both the customer and the company. Such solution spaces almost always win the support of the compliance and legal departments because it circumscribes broad limits but allows tailoring the solution to the individual customer circumstances. The employee can break the rules without breaking the rules.

Ron Zemke, who authored a number of landmark books on customer service, found that when employees were given five to ten sets of benchmark guidelines, they felt comfortable extrapolating to other solutions as needed. This willingness to take such risks is contingent upon clear reinforcement from supervisors and executives.

The empowerment policies should be presented in a four-part format. The first section is the basic response and the so-called standard solution. The second part is the range of types of factors that might affect the resolution and examples of the types of limits on the resolution or remedy. When needed, guidance on how to take the actions implied by the resolution or remedy should also be included. A key point is that neither the range of factors nor the limits of the resolution are presented as definitive or complete. They are presented as examples.

The third section is a listing of the probable challenges and arguments the customer may bring to the situation, such as, "How could this have happened?" or "Why was I not told this?" or "Do you have a known defect in this part of the car?" The employee should be armed with the best clear and believable response to each anticipated challenge. The final section should provide the employee with background and reference materials, as well as a list of internal contacts for when customers wish to escalate their challenge.

Information on the Customer's Situation

Flexible guidance is useless in applying empowerment if you lack the context of the customer's request. Technology, especially CRM and KMS information, should be immediately available to the frontline employees.

Time to Create Connection

The time needed to connect with a customer is one of the most uncontrollable of the factors that employees and management face. For example, when bad weather results in five flights being cancelled in a 20-minute time span, the six airline gate staff will not have time for in-depth empathizing and connecting. If some of the ramp workers and other airline back office staff are mobilized (and have already been trained to rebook customers), the six employees could become 20. Deciding when there is available time to create a connection is one of the key areas of employee discretion. After making this point, I recently asked BC Johnson, engagement manager of The Disney Institute, how frontline cast members can create a connection with each of the thousands of guests streaming in the front gate on the morning of a busy day. He indicated that employees are expected to be enthusiastic and genuine to more than a dozen guests a minute and are trained to do that for hours on end. It takes a special person to maintain that level of verve for that long.

Reinforcement

Employee confidence in their empowerment and skill in connecting with customers is enhanced by continuous reinforcement of both the employee's specific actions and their willingness to take a risk and try new approaches. This reinforcement should be done right after the transaction via feedback from supervisors and peers, as well in periodic team sessions.

At Blinds.com, a full-service window blind company, supervisors have a goal of reinforcing every employee's behavior at least once a day. Supervisors listen to one or more calls via monitoring and make at least one positive face-to-face comment to the employee about some aspect of the call. Further, while walking among the group, any time they overhear an employee's adept handling of a customer, they mention it. And all these comments are made in front of the group. Finally, they conduct weekly drills for skills reinforcement training. This training focuses on one of a dozen specific skill sets, either one that is flagged as an area of weakness by recent evaluations or one that has not been addressed in the last three months—even the best staff get sloppy without reinforcement.

A best practice for reinforcement is to hold what I call victory sessions. In these sessions, held once every two weeks, every team member talks for 60–90 seconds about the toughest situation they *successfully* handled in the last two weeks. Each employee receives peer and supervisor recognition for a job well done. Also, everyone receives education and ideas on how to handle similar situations. Further, empowerment and risk taking are recognized and celebrated, reinforcing both behaviors. Finally, it creates the incentive to go the extra mile for customers when a truly difficult customer is encountered; employees know that their successful response to the situation will make a good story for the next team victory session. One company is considering video-recording such sessions and putting selected stories on the company intranet; twenty-somethings love seeing themselves on YouTube-type media.

Planning for Emotional Connection

The company must plan to create opportunities for connection. The purchase of a few items in a supermarket or online may not lend itself to connection. Even when the checkout in the supermarket is with a human cashier rather than a scanner, connection may be difficult. In many supermarkets, the cashier is told to make eye contact and give a friendly greeting and ask whether customers found everything they were looking for. In most cases, they are instructed not to start a conversation that will slow the checkout process or create distractions that will lead to keying errors. The cashier must be trained to read the situation.

Planning for Connection

There are two strategies for planning a connection. The first, like the earlier PetSmart example, is to create a number of inherently high-involvement transactions that are likely, if executed well, to provide an opportunity for connection. If an owner brings her dog in for grooming or doggy day care and the employee relates well to the pet, as well as to the owner, there is a high probability of a positive connection. Further, if the employee sends two emails during the day-care period with a picture of the dog cavorting with its new friends, connection is almost certain.

The second approach is to take a plain vanilla transaction that would not normally be memorable and make it memorable. For this to work, the employee must either spend enough time engaging with the customer or have enough information to execute one of the following connection approaches. The first strategy is more natural. The second is much more opportunistic and depends on how well the employee reads the customer.

Some customers do not want connection, maybe because of their personalities, maybe because of circumstances like being in a rush. The employee must first read the customer and then decide whether connection is appropriate. Also, if the customer is on a long line, providing timely service to everyone may trump creating connection with every customer. The individual employee must decide whether connection is appropriate both from a time investment and customer desire perspective.

Six Approaches to Connection

Connection can be created in at least six ways. Several require only a small amount of time and creativity but produce significant lifts in loyalty—often more than a 20 percent increase, as noted in Chapter 1. Even creativity can be taught. Many of these methods are predictable, but even those that are situation specific can often be planned in advance and executed electronically.

Take the High Road by Placing the Customer First

When a company discourages the customer from making a lucrative but inappropriate purchase, the emotional impact is powerful. When, for example, an investment advisor suggested the customer use the money to pay off his credit card bills rather than buying an investment under discussion, that company gained lifelong loyalty. As the customer noted in a subsequent survey, "While I was angry at first, I then understood that the company was more interested in my well-being than in making a quick sale—I'll be a customer for life!"

Identify with the Customer

Customers identify with salespeople who are similar to them in some way—fellow pet owners, neighbors, or fans of the same sports team. The one challenge for salespeople is how they reveal these common interests; they need to show they know who customers are without being too invasive or familiar. The ideal is to be friendly but not creepy.

Anticipate Customer Needs

A good example of anticipating customers' needs is at the Automobile Club of Southern California (Club). The Club has a Water Program that anticipates customer needs. First, knowing that every minute spent waiting for the tow truck while broken down on the freeway with cars whizzing past feels like five minutes, the Club provides progress reports via technology on how soon the tow truck will arrive. Upon arrival, the towing driver goes to the driver's window and apologizes for being late, even when the truck is early, and then hands the customer a cold bottle of water. This show of concern and anticipation often converts satisfied customers into delighted customers.

Friendly Interaction

American Family Insurance provides a great example of a friendly interaction. The insurance agents have data showing who is a veteran, and on Veteran's Day they call or email their customers to thank them for their service. Policyholders and even their families email back, thanking them for the recognition. In a different example, an acquaintance always goes to a café for her morning coffee with her dog. Her local Starbucks has a bowl of doggie treats, and the dog always gets a treat while his owner pays four dollars for her coffee. The customer says the treat for her dog assures that she will never go elsewhere for her coffee. You can do the cost-benefit on the treat expense versus the coffee revenue.

Education and Demonstration of Expertise on the Topic

One of the most powerful delighters is when an employee takes extra time beyond the basic transaction to educate the customer on how to get more out of a product or how to avoid a potential common problem. The company receives four benefits from encouraging such activities.

First, customers are delighted that the employee has taken extra time to assist and educate them. Second, the company benefits because the customer either gets extra value and feels less price sensitive or avoids a problem. Third, the employee feels good about going beyond the basic robotic transaction to add value for customers. Finally, when staff members show expertise in the product area, customers believe that the company's products must be the best because the employees are such experts. A two-minute investment in these activities can return a 10–30 percent increase in the highest rating on loyalty surveys.

Financial or Value-Add

An example of this approach is the O'Hare Airport Starbucks. Because I had to wait for two minutes to get fresh coffee, the employee did not charge me for it. Special discounts are effective delighters but can then become expected in subsequent similar situations, which is a liability. A second approach is to provide expertise and extra value. One personal products company gave teenage girls a guide on how to be great babysitters—an extra valuable service for their basic target market.

Each of these connection activities provides value but has some cost in terms of staff time. The connection should not be routinized; otherwise, it may become robotic. One of the worst examples of supposed extra service I've ever seen was at a supermarket chain. The store wanted to help elderly customers carry their grocery bags to the car. An executive decided that asking every customer if they needed assistance should be made a standard part of all transactions. In some stores, all customers buying even a single loaf of bread or bag of chips were asked if they needed help, which made both the employees and the company look ridiculous and was insulting to the young construction worker buying two items.

Emotional Connection in a Techno World

The key to balancing connection and technology is to understand that most parts of the DIRFT transaction just need to be efficient and transparent and that only one or two aspects need to provide memorable connection. As more purchases are completed online, opportunities for emotional connection decrease slightly, although concern and connection can still be achieved via chat or email or even by carefully personalized automatic responses.

Creating the actual emotional connection has usually been left up to people but can also be accomplished through technology. Consumers are now used to communicating with each other via technology, but they are happy and impressed by human-like interactions that create a connection. A few examples will illustrate:

- Interacting with someone by email can be humanized with a photograph.

- One chemical company provides both pictures and factoids about their employees, such as "dog-lover."

- Some companies have used smiley faces and frowns to convey emotion.

- A cellular phone company encourages its employees to mirror customers' demeanor and language on the phone and in their stores.

Again, the challenge is to be familiar and human without making customers feel nervous and think that their every move on the Web is being monitored.

The Role of Executives and Supervisors

Leadership will make or break any culture. Executives set the direction and tone, but supervisors, who are the sergeants of the CE army, ensure the actual delivery of the CE. Both groups are critical and must be aligned.

Executive Actions Needed to Create a Customer-Focused Culture

To foster a culture of empowerment and connection, executives must take four specific actions: emphasize CE in communications, walk the talk, make CE a key attribute for appraisals and incentives, and break down barriers to frontline success.

Emphasizing CE in Communications

When executives communicate, they tell everyone what is important by the order in which they address issues. Dan Hesse, CEO of Sprint, noted that in quarterly financial analyst calls, he talks about customer satisfaction before financial results. Fred Smith, Founder, Chairman, President, and CEO of FedEx, stresses the Purple Promise: "Make every FedEx experience outstanding." A financial services company features frontline customer service staff throughout its annual report. Such actions show employees and customers that the experience is paramount. At the Pointe Hilton, monthly all-hands meetings celebrate great service transactions as reported by surveys and compliments or as observed by team members.

Walking the Talk

At Disney, everyone, including executives, is expected to interact with guests and to pick up a piece of trash if they see it. At the Pointe Hilton, every executive must spend time in the lobby interacting with customers as well as inspecting suites for cleanliness and calling both recent arrivals and survey respondents. It is critical for frontline employees to see that executives stop their "important duties" to deal with a customer who has an issue.

Making CE a Key Part of Accountability, Evaluation, and Incentives

Putting service, empowerment, connection, anticipation/prevention, and VOC input into the job description is a prerequisite to establishing accountability. If it is not explicit, it does not exist. Blinds.com has a great approach to accountability, evaluation, and incentives. The job description and evaluation sheets have two parts. One part is objective, stating what behaviors are expected. The second part addresses how well the behavior is performed. My point is that all the behaviors must be in the job description, but the nuances can be left to the evaluation and incentive process.

The most important accountability behavior is for frontline supervisors to support the five frontline roles. Executives should review the job descriptions of frontline supervisors to ascertain whether a culture of service (whether to internal or external customers) is being fostered. If it is not supported at the supervisory level, it will not happen!

Breaking Down the Barriers to Frontline Success

Marisa Mayer, the CEO of Yahoo, talked in a recent speech about "getting out of the way" so that employees could be successful. In pursuit of this goal, she had created a PB&J internal website that asks employees to highlight Process, Barriers, and Jams that are impeding their progress. If more than 100 employees vote for an issue, management commits to fix it quickly. Frontline employees have been convinced that anything can be changed if it is getting in the way of better serving customers and company success.

Supervisors' Role in Creating a CE Culture

Supervisors are the critical factor in the success of empowerment and the creation of connection. If supervisors do not actively support empowerment in terms of delegating, coaching, motivating and listening, and soliciting input, these things will not occur.

Delegating Authority

Delegating is very hard for a supervisor but creates a huge win-win-win if it is done correctly. If delegation works, the employee and customer are happy with an efficient, satisfying transaction, and the supervisor saves the time of involvement.

The reason supervisors do not delegate is that they do not trust either the employee or the customer, neither of which makes sense. If they do not trust the employee, they fear that the employee will give too little or too much. If the employee is trained in satisfying the customer, giving too little will not happen, and it is almost impossible to give too much. The worst the employee will do is create a delighted customer. If supervisors do not trust customers, why punish the honest 98 percent in order to catch the 2 percent who are dishonest?

Michael Ellis, director of rooms at the Pointe Hilton, says the toughest action a front desk staff member must take is telling guests with valid reservations that the hotel is overbooked and they must go to another hotel at the Hilton's expense. Dealing with a tired, angry customer is difficult. Getting the front desk supervisor to allow staff to handle this situation themselves when they feel equipped and ready is even more difficult. Ellis counsels supervisors to encourage staff members to handle the situation themselves, with the knowledge that the supervisor is observing in the back room via a video camera and will immediately rescue them if they need help. Once an employee has successfully handled this most difficult situation, the employee's self-confidence and the supervisor's trust in the employee soar.

Coaching

Chris Blair of Blinds.com makes the savvy observation, "Telling is not coaching." If you tell an employee what you think, you are not supporting self-discovery, and you are not coaching. You could say, "What do you think you did well, and if you had it to do over again, would there be anything you would do differently?" Then you are teaching your employee how to practice self-improvement. After the employee has answered, provide feedback. The supervisor should provide coaching and feedback based on the immediately preceding service interaction, rather than hours later or, worse, in weekly or monthly feedback. This feedback, which stresses a high ratio of positive over negative feedback, should be done in front of other team members so that there is peer recognition as well as learning.

Recognition

Research by Gallup Inc. and my own experience suggest that recognition or the lack of it from supervisors is the top reason for turnover among good employees. What is ironic is that recognition costs almost nothing to provide. Even the lowest performers are doing their jobs well most of the time. Great supervisors make recognition a priority for both daily individual public feedback and group meetings, such as the victory sessions described earlier in the section on "Reinforcement."

Listening and Input

Gathering and passing up feedback provided by the frontline staff on company process and policy has a major impact on employee morale and is seldom addressed. For gathering feedback to be useful, the supervisor must tell the frontline staff what is being done with the feedback. When this process works well, employees have a much stronger sense of control over their environment and much lower frustration levels. This activity should be conducted in parallel with both employee surveys and direct input channels, such as Yahoo's PB&J website.

Metrics for Measuring and Managing Empowerment and Connection

The two most prevalent and obvious indicators of empowerment and connection are customer feedback (both complaints and compliments) and escalation of transactions. However, there are other good metrics for measuring and managing empowerment and connection. As with other parts of experience measurement, multiple data sources provide a more valid picture of the situation.

Customer Feedback

Customer feedback is the most obvious source of information on empowerment and connection. Customer surveys and complaints often indicate that employees either did not use empowerment or failed to effectively connect. On the other hand, compliments often indicate when employees took the initiative to push the envelope or be creative in order to solve a problem. Two survey questions that are useful for measuring connection are:

- Did the staff member genuinely care about your issue? (Yes ... Somewhat ... No)

- What one word would you use to describe our company? (Open-ended answer)

Another approach is to place the message, "How am I doing? Tell my boss!" in the signature block of emails to internal and external customers, along with the supervisor's email address. Surprisingly, more positive than negative feedback is usually evoked.

Empowered Action for Difficult Issues

The percentage of transactions in which employees take action to resolve the problem on first contact is a good indicator of empowerment. Ninety percent of interactions with customers are straightforward; empowerment manifests itself in the other 10 percent. Therefore, when identifying interactions to review or observe, the supervisor should strive to find the exceptional situations. These cases can be identified by using the reason for contact when drawing the sample from the CRM system. However, routine transactions are also opportunities for connection, so there should be at least some review of how often connection is attempted even during simple transactions (remember that this can include preventive education).

Escalations to Supervisors or Other Departments

With proper empowerment, few if any transactions get escalated. Unnecessary escalations are a useful process metric of empowerment but can be misleading. Most escalations to a supervisor are indicators of employees' lack of empowerment or lack of confidence in their skills to use it. But the lack of escalated complaints does not guarantee that employees are using their empowerment effectively. This is because employees can simply say no to customers or tell them what they want is not possible and convince them to give up. Therefore, no escalations does not guarantee the effective use of empowerment unless you can also observe customer contacts or survey customers to confirm that complaints are actually being handled correctly.

One useful approach is to require employees to indicate why the employee is escalating a contact—for example, the lack of authority, skill, or information or a customer's refusal to accept the employee's response. The last could indicate either a communications training issue or an unreasonable customer.

Monitoring and Observation

Monitoring of telephone and email/chat conversations should focus on fully using empowerment and taking appropriate advantage of opportunities for connection. Technology can assist this activity via speech and text analytics.

GETTING STARTED

Questions to Ask Yourself About Your Culture of Empowerment

1. Does the frontline job description include connection and prevention as responsibilities?

2. Does the frontline staff have the empowerment, as well as the information, to resolve at least 95 percent of all the issues they encounter?

3. Are supervisors mandated to encourage empowerment and connection and accountable for doing so?

4. Do executives emphasize empowerment and connection, and do they walk the talk?

5. Does the company have the tools to measure empowerment and connection at the individual and system levels?

KEY TAKEAWAYS

- A culture of customer focus and emotional connection cannot happen without empowerment that allows the frontline staff to handle at least 95 percent of issues. Empowerment requires flexible response rules supported by easily accessed information.

- Connection can be created in at least six ways, most of which have little or no cost except for the investment of frontline staff time but all of which can significantly boost customer loyalty.

- Executives must signal their strong support for empowerment and connection, and supervisors must be trained to actively support connection recognition and delegation of authority.

- The frontline staff must be trained on how to connect and how to use empowerment.

- Companies must measure empowerment and connection actions if they want to manage these activities.

Notes

1. David Lenhardt, CEO of PetSmart, "Barking Up the Right Tree: How Services Can Help Differentiate a Business During Bad Economic Times ... and Good," presentation to Center for Services Leadership, Arizona State University, November 7, 2013.

2. Roger Connors and Tom Smith, *Change the Culture, Change the Game* (New York: Portfolio/Penguin, 2011), p. 33.

CHAPTER 10

Leading the Charge
to the Next Level

Recently, I led a seminar for CE and service strategy executives and asked the participants about their greatest challenges. To my surprise, a strong minority said the survival of the CE function was their chief concern. Most facing a challenge to the existence of their operations said the CE had been made responsible for everything related to the customer. CE executives who were struggling said they had difficulty demonstrating the economic impact of their activities and had alienated other important executives too.

However, many CE leaders do flourish and get promoted to very senior positions. This chapter will:

- Explain the roles of successful CE leaders, summarizing their most important functions.

- Describe two CE roles fraught with danger that should be avoided.

- Describe the lessons from six organizations' CE journey.

- Prepare you for the financial and organizational challenges that will arise.

Understanding the Role of the CE Leader

The CE leader should ideally be at an executive-level position but often is not. This is especially true when CE leaders are selected from within an organization. The CE leader must have specifically defined functions to allow the executive committee to understand the CE leader's contribution. If the job sounds like fluff, it will be treated as fluff.

Following are descriptions of the seven appropriate and necessary CE leader functions and descriptions of two functions that CE leaders sometimes assume but should not. These two functions are inappropriate for the CE leader, and taking them on often leads to failure.

Oversee the Mapping and Analysis of the Current CE and Ideal CE Process

The CE leader does not need to be an expert in process mapping but should understand it enough to manage the mapping exercise. Corporate process improvement staff should do the actual mapping. (How to conduct and analyze process mapping was outlined in Chapter 4 and described in more detail in Chapter 8 in terms of the application of technology.) The current process is mapped, and then the ideal process is suggested along with intermediate stages.

Gather Unified VOC Data on CE

Ideally the CE leader should at least coordinate the overall VOC and own the VOC data collection planning process (described in Chapter 7) to ensure that the information collected from the different sources will fit together to produce a single, unified picture of the CE. This analysis will also identify which fixes or changes will have the greatest impact on overall satisfaction.

Create the Economic Imperative for Action

Creating the economic imperative for action means analyzing the data collected for the unified VOC. This economic imperative is predicated on quantifying the value of the average customer, the prevalence and impact of problems encountered by that customer, and understanding how problems contribute to the attrition of customers and lost revenue. (This analysis was described in detail in Chapter 3.)

Facilitate Identification of Opportunities

Executive management identifies priority problems and opportunities that should be addressed. However, the CE leader must take the initiative in presenting the options and creating the economic imperative for priority attention, using the analysis in the next function. Russ Fleming, VP, Strategy, Innovation & Product Development at FedEx, suggests two actions that get executive attention: "First, use video snippets of real customers telling about real experiences. Second, quantify the estimated revenue implications of customer points of pain in a manner finance finds credible. The combination of humanizing and quantifying the issue's impact assures executive attention."

Facilitate Action Planning

Action planning requires assembling the midlevel managers and frontline staff who know how the current process actually works and challenging them to identify not only the needed fixes but also how to make those fixes. This activity is not a single one- or two-hour meeting but most likely several longer meetings. If the problem being addressed had a simple fix, it most likely would have been fixed already.

Fixes can include resetting customer expectations or proactively educating the customer. All supporting functions, such as IT and human resources, should also be present as well as any important partners such as field staff or dealers who will be affected. This activity also includes the establishment of metrics to track goal achievement and suggestions on how much movement should be expected from the recommended actions. For example, if a process is changed, does the action team expect that all customer calls about a particular invoicing error will be eliminated, or is a 50 percent decrease a reasonable outcome to expect?

Finally, as most actions will require cross-functional coordination, the action planning group should reach consensus as to which groups should take the lead and have primary responsibility for executing the plan. If the group cannot reach consensus, the CE leader should suggest who needs to take the lead in addressing the problems or encourage a functional leader to volunteer to take the lead. The responsibilities for the multiple initiatives should be divided so that one or two executives are not overly burdened.

Measure and Celebrate Progress

As noted in Chapter 7, many VOC processes are never systematically monitored to determine whether the plans made in response to the VOC reports ever achieve the promised improvements. As W. Edwards Deming said, "You can expect what you inspect."[1] The lack of accountability is a serious impediment in achieving action, and being accountable by measuring progress is one of the most important functions the CE leader can perform. Another key part of this process is recognizing and celebrating successes and ensuring all the involved actors receive accolades. If anything, spread the glory too wide; then those who were not strong contributors on the current project will work harder next time.

Act as an Advocate for Customers to Top Management

The CE leader must tell management the bad news along with how much money management is leaving on the table by not acting. My experience is that executives appreciate an objective opinion of what is working and what is not working from a perspective outside the operational chain of command. In doing this, you must be sure not to blindside any executive. At one auto firm, the CE leader encouraged the functional heads to present both the good news and the bad news to the COO, and the CE leader confirmed what was said. However, the functional executives must be aware that if they sugarcoat an issue, they will get caught in the long term by the objective VOC process.

Two CE Roles Fraught with Danger

The CE leader should never accept two functions because they are impossible for an executive external to the operating chain of command to achieve: (1) assuming total responsibility for customer satisfaction and loyalty and (2) fixing all quality and service issues.

Owning All Corporate Customer Satisfaction

Customer satisfaction and loyalty (as Chapters 2, 3, and 4 have shown) are the result of product and marketing decisions as much as CE. If the CE leader becomes solely responsible for improving the corporate customer satisfaction indices, the operating and functional executives are off the hook. Once accurate data is available, the impact of CE initiatives can be discerned and separated from other market factors.

Fixing all Quality and Service Issues

When it comes to fixing quality and service issues, the CE leader is acting as a consultant for the rest of the executive team, composed of the operating and functional executives. The operating and functional executives are the leaders who should be assigned responsibility for achieving improvement. Even if the issue is cross-functional, a single executive should be made responsible for leading each improvement effort. That executive should not be the CE leader but rather whoever must obtain cooperation from the other groups involved. At one very successful company, the CE leader had identified 20 opportunities across the company, and his name was associated with only two of them. The other 18 opportunities were assigned to other operational and functional executives. The CE leader reported to the COO on progress via a monthly status report.

Lessons from the Journey

The single truism of every CE journey in an organization is that there will be ups and downs. As with customer experience, the company's

CE journey will occasionally have unpleasant surprises. The key is to prepare yourself and your management to deal with those surprises.

This section relates the challenges and keys to success found in six organizations as they established or improved the CE process. These companies all went through similar phases in their multiyear journey.

The six organizations from which the following lessons are drawn are (1) a restaurant chain, (2) a vehicle finance company whose customers were retail dealers, (3) an auto manufacturer, (4) a travel and leisure company, (5) a business services logistics company, and (6) a technology company. Each company experienced most but not all of the stages described here. However, they all experienced both economic and organizational upheaval. The economic upheaval included a financial crisis leading to reduced demand and bankruptcy (in one case), the entry of a new competitor or product, and/or a serious quality gaff that led to bad publicity and regulatory involvement. The organizational upheaval included a new CEO, a merger or acquisition, or a serious reorganization that led to employees worrying about their jobs.

We Have a Problem!

In most cases, a call to action came from a top executive, often as the result of abysmal business results. At one company, a new CEO pounded the table and insisted, "This cannot continue!" In two cases, management was happy with existing results but feared that rapid growth would dilute the current good CE. In one instance, a midlevel manager in the service area saw and presented an opportunity for improvement with a satisfaction and bottom-line payoff.

In almost every case, a key to getting executive commitment was providing case studies of actual customer situations. Several companies found that videos of customers telling their stories were very effective. Further, the business services company required their executives to work on the front line interacting with customers and helping customers use the service. In this way, the executives learned about the sometimes inconvenient processes that the company asked customers and frontline employees to execute.

Who Is Going to Fix this Situation?

The CEO or the executive sponsor usually presented the customer service problem to the entire management team. However, a midlevel manager in marketing, quality, or service was usually tasked with solving the problem. Very seldom was a senior executive immediately placed in charge. In many cases, the designated manager was told to take three months and come back with a plan. Seldom was this position thought of as a permanent position; it was more that of a task force chair. Only after the benefit of a continuous improvement process was identified was a permanent CE leader position created.

Gather Data

The newly appointed CE leaders used two very different approaches to data collection. Two of the companies gathered anecdotal data, and four of the companies executed a comprehensive assessment of the CE using internal and external data. Although the anecdotal approach was faster and cheaper, it tended to highlight mainly operational issues that were low-hanging fruit. The second approach, while taking longer and costing more, often highlighted broader issues such as fundamental product problems, customer expectations, and delivery process issues that were missed in the more cursory approach. The research also quantitatively highlighted key drivers of satisfaction that would give the greatest leverage in improving overall satisfaction.

When companies gathered anecdotal data, overall satisfaction rose, but then improvement stalled because the more insidious institutional issues had never been identified, let alone addressed. When companies did a comprehensive assessment and identified and focused on the fundamental drivers of satisfaction, improvement continued, though it slowed down after the first two years. Sustained, annual research kept the continued focus on key satisfaction drivers.

Collecting data on the cost and revenue impact the customer issues (as outlined in Chapters 3 and 7) was critical no matter which approach was used. When the finance department could not provide data on the value of a customer, the CE leaders used a painfully low estimate of that value to estimate revenue damage. They used operations data to estimate how many customers were impacted and survey data to estimate the impact on loyalty. Combining the three sets of data (value of customer, problem incidence, and damage to loyalty) allowed the calculation of the number of customers at risk and the revenue impact of the issues. In each case, the business case was compelling enough to keep top management engaged and supportive. In two cases, the finance department revised the very conservative estimate of the value of the customer to arrive at a significantly higher impact of the problems of the status quo.

The companies also considered the cost of winning a new customer, which is almost always much more than the cost of keeping a customer by turning around a negative situation. In B2B environments with long-term contracts, this calculation is less persuasive, though share of business is a very powerful metric. The financial services company found that the recent CE was a strong positive driver for the bulk of their customers' future business. A final type of analysis that was very effective was comparing the cost of handling a problem to the cost of preventing it from occurring.

Organizations were more effective in capturing granular, actionable operational data on problem occurrence and on the overall CE when an existing IT platform was used to run the basic product delivery or when the sale-to-cash operation was modified to allow the collection of granular data on operational defects that could then be tied back to the available customer feedback. In the auto company, the operational system platform was the warranty database; in the restaurant chain and the business services company, the system platform was the order/billing tracking system. Likewise, the reservation system was used to collect CE data in the travel and showed refunds for unsatisfactory rooms or meals. that resulted in a credit or refund was easily enllowing a unified analysis of CE, operations, neters.

blogy and financial services companies did not ation management systems that could easily log operational exceptions. These technology and financial service or-ganizations were forced to create new systems to track customer problems and the descriptors of CE quality. If the IT department is made aware of the value of such a capability, the next generation of operational systems can often be tailored to support the CE manage-ment function.

The advantage of modifying and using an existing system to gather problem occurrence and overall CE data can be illustrated with the restaurant chain's approach. The server order entry/billing system feeds directly into the kitchen production process, which notes every employee who touches the order. When the server encounters either a customer problem or a compliment, this information is en-tered into the system. The server will enter one of more than 20 codes describing the problem, supplemented by a verbatim statement by the customer of the issue and an indication of whether the customer was given a free dessert or the meal was removed from the bill. The existence of the issue and its detailed description, including the dish (e.g., hamburger ordered medium rare was cooked medium), can be tied back to the cook and the kitchen manager. Also, if there is a sys-tematic complaint, such as a meal being reported as too salty or too spicy, a quick analysis can be performed on whether it is a locational, regional, or national problem.

At the restaurant company, granular data on the CE of all 80 mil-lion meals served annually can be tied back to individual products and employees, thereby creating strong accountability as well as the ability to quickly monitor the impact of process improvements. It is impor-tant to note that the servers in this chain do not just field complaints verbalized by customers. They are trained both to build rapport and to examine the plates when they clear the table. Meals may be half-eaten because the customers were satiated and wanted to take the rest home or because they were not really pleased with the food. The server is trained to ask whether the customer wants to take the left-over meal home; if not, the next question focuses on whether the cus-tomer was satisfied and on a possible remedy.

Action Planning

All six of the companies have an action planning protocol for addressing opportunities. This protocol must be formal and, for cross-functional issues, must include all major functional areas of the company. Once the key priorities have been established by executive management, a cross-functional team uses the priorities to brainstorm specific actions that will optimize the impact on the customer. This brainstorming will start with a large number of quickly developed ideas (ideally from multiple cross-functional groups) and then distill the many ideas down to three to five priorities across the company as a whole. Additionally, each department or geography will have action planning sessions that address priorities in their area.

For the corporate-level action planning to be successful, an owner of the plan must be identified. In addition, an action plan, schedule, resource requirement statement, and process metrics should be articulated. An additional aspect that can be very helpful is to conservatively estimate the number of problems that will be prevented if the action plan is implemented—and then be sure to track the problems. These action plans will be submitted via the CE leader to executive management, and progress reports will be submitted monthly or quarterly. For example, in the business services company, invoice adjustments were an expensive dissatisfier. The company established the parameter of invoice adjustment calls and developed an estimate of the cost per call. When calls declined, the finance department immediately gave CE credit for cost savings, even before the survey showed happier, more loyal customers. Ideally all actions can be completed within six months.

Impact Tracking

As noted in previous chapters, accountability is critical to positive CE impact: What gets measured gets attention. Therefore, process measurement should be in place to give monthly feedback on the impact of improvements (ideally weekly, which is what the business services company and restaurant chain used). Additionally, a periodic customer satisfaction survey should be conducted quarterly or annually, depending on the number of customers being served. For instance, the vehicle finance company had less than one thousand customers, so frequent surveys would not have been cost-effective. This customer survey should not only obtain data on broad satisfaction and loyalty but also capture problem data. In addition, data on satisfaction with key corporate activities should be collected so that process improvements can be tied to satisfaction with the activity.

Celebrate and Move On

Celebration should be consciously directed at three separate audiences: the action team, all other company employees, and customers. The fastest way to win support is to give lavish kudos to the action teams and their leaders. This creates the incentive for other employees to volunteer for the next set of teams. Jeanne Bliss calls this the Tom Sawyer approach to getting others to do the heavy lifting.[2] This same approach of sharing credit is also a standard practice of effective consultants—make your client look good—and the CE leader is primarily an internal consultant. As noted in Chapter 7, it is critical to feed back the results of the action plans to the employee base as a whole and to customers. Not only does this show that the company cares and is improving, it creates additional pride among the action plan team members.

Preparing for the Inevitable Financial and Organizational Upheavals

A perennial problem in many companies is that CE and customer focus flourish in times of growth and profits and then come to a screeching halt when financial disasters or organizational upheavals occur or when CE metrics do not increase predictably. The CE leader must prepare for such events and also educate management that CE metrics do not always reliably increase year over year.

estimate of that value to estimate revenue damage. They used operations data to estimate how many customers were impacted and survey data to estimate the impact on loyalty. Combining the three sets of data (value of customer, problem incidence, and damage to loyalty) allowed the calculation of the number of customers at risk and the revenue impact of the issues. In each case, the business case was compelling enough to keep top management engaged and supportive. In two cases, the finance department revised the very conservative estimate of the value of the customer to arrive at a significantly higher impact of the problems of the status quo.

The companies also considered the cost of winning a new customer, which is almost always much more than the cost of keeping a customer by turning around a negative situation. In B2B environments with long-term contracts, this calculation is less persuasive, though share of business is a very powerful metric. The financial services company found that the recent CE was a strong positive driver for the bulk of their customers' future business. A final type of analysis that was very effective was comparing the cost of handling a problem to the cost of preventing it from occurring.

Organizations were more effective in capturing granular, actionable operational data on problem occurrence and on the overall CE when an existing IT platform was used to run the basic product delivery or when the sale-to-cash operation was modified to allow the collection of granular data on operational defects that could then be tied back to the available customer feedback. In the auto company, the operational system platform was the warranty database; in the restaurant chain and the business services company, the system platform was the order/billing tracking system. Likewise, the reservation/billing system was used to collect CE data in the travel and leisure company; it showed refunds for unsatisfactory rooms or meals. Any customer issue that resulted in a credit or refund was easily entered into the database, allowing a unified analysis of CE, operations, and revenue/cost parameters.

Oddly, the technology and financial services companies did not have overall information management systems that could easily log operational exceptions. These technology and financial service organizations were forced to create new systems to track customer problems and the descriptors of CE quality. If the IT department is made aware of the value of such a capability, the next generation of operational systems can often be tailored to support the CE management function.

The advantage of modifying and using an existing system to gather problem occurrence and overall CE data can be illustrated with the restaurant chain's approach. The server order entry/billing system feeds directly into the kitchen production process, which notes every employee who touches the order. When the server encounters either a customer problem or a compliment, this information is entered into the system. The server will enter one of more than 20 codes describing the problem, supplemented by a verbatim statement by the customer of the issue and an indication of whether the customer was given a free dessert or the meal was removed from the bill. The existence of the issue and its detailed description, including the dish (e.g., hamburger ordered medium rare was cooked medium), can be tied back to the cook and the kitchen manager. Also, if there is a systematic complaint, such as a meal being reported as too salty or too spicy, a quick analysis can be performed on whether it is a locational, regional, or national problem.

At the restaurant company, granular data on the CE of all 80 million meals served annually can be tied back to individual products and employees, thereby creating strong accountability as well as the ability to quickly monitor the impact of process improvements. It is important to note that the servers in this chain do not just field complaints verbalized by customers. They are trained both to build rapport and to examine the plates when they clear the table. Meals may be half-eaten because the customers were satiated and wanted to take the rest home or because they were not really pleased with the food. The server is trained to ask whether the customer wants to take the leftover meal home; if not, the next question focuses on whether the customer was satisfied and on a possible remedy.

Figure 10-1

Trend in Customer Loyalty to and Satisfaction with a Customer-Focused Company (Actual Loyalty and Satisfaction Data, Indexed)

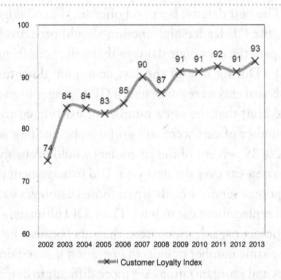

With many externally caused events impacting any organization, macro customer satisfaction and loyalty metrics often do not increase in a straight line. For instance, the data in Figure 10-1 describes the customer satisfaction trend from one of the six previously described companies. The data shows strong increases in the first year, then plateauing, and then an upward but bumpy trend. All six of the companies have experienced similar trends, and one of them recently won the Malcomb Baldrige National Quality Award.

Disaster can come as either economic or organizational upheaval, and each has its specific challenges for the CE leader. Fortunately, there are approaches that have been successful in protecting the CE initiative from both types of disasters.

Economic Upheaval

Economic upheaval includes economic downturn, quality crisis, a merger with another large organization (also an organizational upheaval), the sale of the company, or a new and serious competitor or product. In each of these situations, a standard first reaction is to conserve cash, cut costs, and stop taking risks.

Here, my advice is the same as in dealing with potential customer problems: The best defense is a good offense. When budget cuts are threatened, the CE leadership function should proactively show, if cuts do happen, the economic damage that will result from the deteriorated CE. During one of the last economic downturns, when across-the-board cuts were suggested, a CE manager in an auto company pointed out that the same number of existing customers with the same number of cars were still going to be needing service and that about 20–25 percent of the customers would probably be in the market for a new car over the next year. Did management really want to provide poorer service exactly when those customers were considering which replacement car to buy? The COO ultimately exempted customer contact organizations from the cuts because, he said, "We still have the same number of units in operation that require service."

Mergers and reorganizations are more difficult to defend against. The best defense in these situations is to highlight the competitive benefits of the CE improvement initiatives underway. Quantify this impact using conservative estimates of the number of customers that will be negatively impacted if the initiatives are halted. Accompany this numerical estimate with a two-minute video of two customers telling their stories. In this way, you appeal to the executives' quantitative and emotional sides.

Economic disasters often come from quality and regulatory disasters. My experience is that many of these disasters take place when companies have reached the pinnacle of service or product quality and become the recognized leaders within their industries or even across all industries. In looking at what happened at seven companies in seven different industries, cost cutting was the underlying cause. In almost every case, I found that there had been at least three turnovers in executive management in the service or quality area. There was no institutional memory as to why many of the quality, feedback, and analysis procedures were in place. When the new management came in, their question was, "How do I make my mark? We're already the top company in quality." The answer appeared to be obvious: cost cutting. Unfortunately, resources were cut in the areas that had assured the excellence of service or quality, leading to deteriorating quality and ultimately the quality, PR, and regulatory disasters.

Organizational Upheaval

The most prevalent organizational changes are the departure of the CEO or of the executive who was most supportive of CE and a major reorganization that leads everyone to stop taking risks and protect turf. Again, the best defense in both cases is exactly the same as in the financial downturn situation: Demonstrate the economic and brand damage that will occur if the CE deteriorates.

Your biggest challenge may be getting access to the new incoming executive. The best approach is to use your state-of-the-CE reporting function, which is probably either monthly or quarterly. Raise the issue of cuts, and then demolish it.

GETTING STARTED
Questions to Ask Yourself About Your Role as a CE Leader

1. Do you have support from at least one senior executive? If not, you will have to prove that you deserve it.

2. Do you have access to process mapping and reengineering expertise? If not, where in the company can you obtain it?

3. Do you have the key data elements to identify an opportunity to impact both the revenue and cost parameters of the CE in a quantified and meaningful manner?

4. Have you constructed a business case for a CE initiative that is compelling to the CFO and at least one senior functional executive?

5. Have you celebrated a CE success where you have lavished credit on the whole team as well as highlighted them to the employee base and the customer base?

6. Are you prepared to defend the CE function against financial or organizational upheaval?

KEY TAKEAWAYS

- The CE leader's primary roles are the facilitation, identification, and quantification of opportunities; supporting goal setting and action planning; and creating accountability among other executives.

- CE leaders should not accept blanket responsibility for all customer satisfaction or for all quality and service issues and indices.

- CE leaders should start by quantifying the revenue at risk overall and then support other executives in achieving one or two victories on acknowledged customer issues.

- The CE journey will be bumpy but will always include the quantification of opportunities, process mapping, target setting, action planning, tracking results, celebration, and organizational and economic upheavals. The CE leader must prepare for and manage all of these events.

- Additional keys to success are the same as for any consultant: storytelling, ideally from the mouths of customers, CFO buy-in, process mapping, and letting the other executives get the credit.

Notes

1. Edwards Deming, "Four Days with W. Edwards Deming," Published Interviews 1986–1993 (Washington, DC: The W. Edwards Deming Institute).
2. Jeanne Bliss, *Chief Customer Officer* (San Francisco: Jossey-Bass, 2006), p. 195.

Index